THE MASTER AND HIS FISH

From the World of *Roderick Haig-Brown*

THE MASTER AND HIS FISH

Edited by
Valerie Haig-Brown

UNIVERSITY OF WASHINGTON PRESS

Seattle

Copyright © 1981 by Ann Haig-Brown

Printed in the United States of America

All rights reserved. No part of this publication may be reproduced or transmitted in any form or by any means, electronic or mechanical, including photocopy, recording, or any information storage or retrieval system, without permission in writing from the publisher.

Library of Congress Cataloging in Publication Data
Haig-Brown, Roderick Langmere Haig, 1908-1976.
   The master and his fish.

   Reprint. Originally published: Toronto : McClelland and Stewart, 1981.
     1. Fly fishing—Addresses, essays, lectures.
2. Fishing—Addresses, essays, lectures. I. Haig-Brown, Valerie. II. Title. III. Title: From the world of Roderick Haig-Brown, the master and his fish.
SH456.H2   1981   799.1'755   81-11674
ISBN 0-295-95847-2   AACR2
ISBN 0-295-95875-8 (pbk.)

Contents

The Man Behind the Rod *9*

Trout
1. In Search of Trout *12*
2. Watch the Creek Mouths *25*
3. The Quinault River *34*

Salmon
4. The Splendour of the Run *44*
5. Diplomat's Fish *56*

Steelhead
6. Along the Steelhead Rivers *63*
7. Steelhead Angling Comes of Age *75*
8. They Pass in the Night *82*
9. Fascinating Challenge *91*

Comparing
10. First Among Equals *98*

Pike
11. Grandmother, What Sharp Teeth You Have! *108*

Other Waters
12. A Westerner Looks at the Beaverkill *115*
13. Chilean Trout Fishing *118*
14. Ever Fish for Sebagos? *125*
15. Salmon of the Vatnsdalsa *135*

    16. On the Trout Water *155*

Flies

    17. Big Floaters for Western Trout *159*
    18. Freedom of Choice *169*
    19. The Evolution of a Steelhead Fly *173*

Ethics and Conservation

    20. Articles of Faith for Good Anglers *179*
    21. Outdoor Ethics *185*
    22. A Talk to Oregon Fly Fishermen *192*

    Acknowledgements *201*

The Man Behind the Rod
(1961)

Fishing is a very flexible sport. You can make of it almost anything you will, from a considerable athletic effort to a quiet mood of observation and introspection. You can even combine these extremes and at the same time enjoy many of the gradations between them, such as delight in the streams and lakes and salt water, joy in recognition of birds and other wildlife or wonder at the beauty of the fish themselves and the intricate mysteries of their ways.

Most of these things are in the sport at any time, but the meanings and emphases one gives them undoubtedly change from time to time. In youth one is likely to be concerned about material aspects of success — a good catch, a big fish, a long cast; one takes pride in bold wading or strenuous travel to a distant place where fish are larger and more abundant. Youth is so vigorous and so aggressive that it has time only for part of the sport and, in truth, the necessary equipment to use only part of it. Youth wants the distant scene, the bold achievement, the obvious satisfactions.

In time, rather quickly as a matter of fact, all this changes. The remote place where unsophisticated fish shoulder each other aside to get at the fly — and there are such places, even today — becomes meaningless as a fishing spot very quickly, though it may still hold the attractions of remoteness; the good catch means nothing, because one no longer goes out with any great desire to make it; even the big fish means less as one comes to know the reasonable and logical limits of possibility. Soon there have been many big fish, all of them much the same size, and many of them less memorable than somewhat smaller ones.

These adjustments are all gain rather than loss, because they leave time for rounded enjoyment of the sport in its true dimensions. It is a large and exciting complex of many things, not simply attention concentrated from two directions upon a hook at the end of a line.

There is no form of fishing that I cannot enjoy at least for a while and by now I have enjoyed most of them. But I may as well admit that I prefer to fish a river or a stream rather than still water. I prefer to be on my feet rather than in a boat. And I want to catch my fish on an artificial fly whenever there is the remotest chance of doing so — which, for trout and salmon, is practically all the time.

I prefer to fish alone and secret unto myself, and usually do so, though I often enjoy fishing with a friend or several friends. If a stream is crowded I will cheerfully turn to its less productive and less crowded parts, or else go away altogether and come back at some less popular time. For I have come out to go fishing, not simply to throw in a hook and pull out a fish.

When I go fishing I want to be a part of the river and all my surroundings, not a stranger thrusting in upon them. I want to move quietly and at my own pace. I want to see and hear and understand. I want to feel that I know something of where the fish are and what they are doing and why. I want to be able to name the birds I see and take time out to watch them. I want to feel the river about me and to fill my mind with the infinity of lights that break from its surface and its depths. I want to know the trees along its banks, the rocks of the bottom and the creatures that shelter there and feed my fish.

In such participation I cannot be a stranger. I belong there as a man should be able to belong in a setting he has chosen for himself. My claim is as good as that of the heron or the otter or the osprey, even though my efficiency may be less.

As for the fishing, it need not be good. There need only be a chance that it may be good. I do not much want to kill fish — I would rather release them. What I want to find is some classic situation — a good fish rising in a favourable, but not too favourable, place; the perfect lie that one can fish down to with mounting concentration, through a long reach; the difficult place close under the bank, where trouble is certain from the log jam above or the rapid below. These things never grow stale.

I may find them or I may not. It matters very little. In searching I shall certainly find other things, expected or unexpected, and from somewhere among them I shall take home at least one bright picture of excitement or beauty to make the day. After forty years, rivers remain places of enchantment and the fish that swim in them creatures of wonder. Some small share in this is the fisherman's real reward.

1
In Search of Trout
(1968)

Fish such as black bass, Atlantic salmon, dorado, bonefish, striped bass, muskies, tarpon, pickerel, or what have you, may be local fishermen's favourites. But for universal popularity and world-wide distribution there is no other fish to compare with the trouts. Trout are practically synonymous with angling, and there are few anglers who do not try for them sooner or later. Many will fish for nothing else during their entire angling lives, and most of these enthusiasts will go as far as funds and time will let them in search of trout fishing that is better or perhaps merely different in some small way from any they have known. If a trout fisherman must stay for a long time in a land where there are no trout, he will almost inevitably look into ways of introducing them; and as often as not he will find some way. Many trout fishermen turn to other fish from time to time, but most are inclined to think of these infidelities as rather questionable adventures — fishing, perhaps, but not really serious fishing.

There are many obvious reasons for the trouts' popularity. Trout are very elegant creatures, clean and graceful in form, usually handsome or even beautiful in colouration. They are vigorous and active in performance, prompt and hearty feeders, and a delicacy on the table. They frequent, by preference, pleasant and beautiful places; charming meadow streams, hill torrents, river estuaries, the shoals and reedy shallows of mountain and lowland lakes. They are to be found at the edges of the sea and two miles above the sea among mountain peaks, in blue waters and brown waters and in waters so clear that the only colours are those of washed gravel and trailing weeds.

In addition to all this the trouts are, for the most part, dwellers in shallow water, usually feeding at or near the surface on a wide variety of fascinating aquatic insects, many of which are themselves very beautiful. They tend to be dainty and selective feeders and they learn quickly to beware of the shadow of man and all the delicate devices he creates for their deception. This naturally leads man on to further extravagances of cunning and ingenuity and so keeps the whole thing going.

The quest of trout started long ago with the brown trout of Europe. Well before the fifteenth century he was recognized as the fly fisher's fish and, with the salmon, something of an aristocrat. Juliana Berners, after declaring the salmon "the moost stately fyssh that ony man maye angle to in fresshe water," complains that though "a gentyll fyssh ... he is comborous for to take." She wastes no time after that: "The troughte for by cause he is a ryght deyntous fyssh and also a ryght feruente-byter we shall speke nexte of hym." Izaak Walton notes that "the Salmon is accounted the king of freshwater-Fish" but calls the trout, without any hesitation, his favourite — "which I love to angle for above any fish." And so the glory grows through the angling works of the seventeenth and eighteenth centuries to the great flowering of the nineteenth century when writer after writer celebrated the trout's glory and precedence, and the sophisticated lore and techniques and equipment of the modern trout fisher took full form and shape.

While the brown trout was attaining his formidable position in the respect and affections of old-world anglers, American settlers were finding and celebrating a trout-like fish in the new world. That it turned out to be a char rather than a trout matters not at all. "No higher praise can be given to a salmonoid than to call it a char." Few fish have more dedicated admirers than has the eastern brook trout, and certainly there is none whose beauty has called forth more resounding praise. A description of the great speckled trout of Lake Edward and Lake Nipigon, wrought by E.T.O. Chambers around the turn of the century, is typical: "Here . . . the American brook trout is found in his most gorgeous apparel. His whole being is aflame with burning passion and nuptial desire, which reveal themselves in the fiery flushes of deepest crimson upon his shapely sides and lower fins. The creamy white margins of the pectoral, anal and ventral fins distinctly mark the course of the fish in the dark water, and form a striking contrast to his olive-colored and vermiculated back and dorsal fin. Here, he has caught the varying tints of the submerged rocks, and of the forest-clad mountains which form the basin of the lake, and in the brilliant brocade of his spotted sides he reflects the gold of the setting sun, and the purple sheen of the distant hills."

So much splendour in a single fish — yet that is the way anglers have felt and written about him time after time, celebrating also his strength and cunning and violent activity on the end of a line. And the brook trout is handsome as few fish are, a lover of clear, cold water and unspoiled streams. If he is less abundant than he once was, it is because he is less tolerant of civilization than other trout and perhaps more responsive to the flies and lures and baits of anglers.

As settlement advanced into the West and finally to the Pacific Coast, two more trouts were discovered — the rainbow and the cutthroat. There are many other fish that are or have been called trout — the lake trout and the Dolly Varden, for instance, both of them chars. But it is on these four — brown, eastern brook, rainbow and cutthroat, with their several subspecies — that the trout fisherman's world is built.

All four have sea-running forms of great beauty and highest performance — the European sea trout, the sea-running eastern brook trout, so abundant in Labrador, the winter and summer steelheads of the Pacific Coast, and the sea-run cutthroat of the Pacific Northwest. Each of these is a special experience, and most trout fishers go in search of them sooner or later. But trout fishing, in all its main traditions and development, is primarily a freshwater sport; it was so to Walton and Cotton, Halford and Skues, Theodore Gordon, George LaBranche, and Edward Hewitt.

The trouts are exacting creatures. They demand abundance of clear, well-oxygenated water; streams with clean gravel for their spawning, rich in insect life for their feeding; summer temperatures generally below 70° Fahrenheit; shade and shelter from the enemies that beset them, under rocks, under banks, or in deep pools. They are at their best in water that is high in mineral content — the chalkstreams of Britain and France, the spring-fed limestone brooks of the eastern United States, the dry-country lakes of British Columbia and Idaho. Yet they thrive also, and often spectacularly, in swamp-water ponds and tea-coloured burns, in the soft-water lakes and rushing creeks of high rainfall areas. They are exacting, yet also highly adaptable.

Of the four, the brown trout is the most cosmopolitan. Long established in Europe, from the British Isles to Yugoslavia and in every country between, he is now a well-assimilated resident in North America from coast to coast. In the southern hemisphere he is happily settled in Chile, Argentina, New Zealand, Australia, South Africa, and probably in other countries with suitable waters. He is the wariest and wisest of the trouts, the most selective in his feeding and the most difficult to deceive by strictly honourable means.

The rainbow is almost as widely distributed, though he has given difficulty at times by disappearing, I suspect because a sea-running stock was introduced rather than a resident freshwater stock. He flourishes close to the equator, on the slopes of Mount Kenya, and twelve thousand feet up in the Andes at Lake Titicaca in Peru and seven thousand feet up at Lake Maule in Chile. He is incomparably the most active and violent of the trouts, as Theodore Gordon recognized long ago in welcoming him to eastern streams. "A better game fish," he wrote, "does not exist . . . rainbow leaps again and again and always runs downstream. Brown and brook trout almost invariably run up, at least their first run is up; but the rainbow, after throwing itself into the air, tears desperately down and you must follow if the fish is any size. Men not accustomed to this trout are apt to lose all the large fish they hook for some time. It fights to the last and when landed has scarcely a kick left in it."

The brook trout has not travelled so far, perhaps because of his preference for colder waters. But he has settled comfortably west of the Rockies and I have found him, large and handsome, in the rivers and lakes of Argentina.

The cutthroat remains the rarest and wildest of the trouts. The hatchery men, I am happy to say, have found him hard to handle; for that reason he has been little transplanted. He flourishes in eastern British Columbia, Idaho, Montana, and Wyoming as the Yellowstone cutthroat, black-spotted, with red-gold sides and a scarlet slash under his jaw. But I admire him most in his sea-running form, when he can be bright as a rainbow and wild as a rainbow, but with a sulky power all his own at the last.

Why do we search for these four fish so hard, so persistently, so obsessively? There are many obvious reasons. Some of us search constantly for trophy fish, those monster trout that make the record books and keep the taxidermists busy. Some search ever for that miraculous river where every cast brings a fine fish to the fly. Some of us are in love with the gear — delicate yet powerful rod, smoothly balanced lines, leaders calibrated to turn over perfectly even into the wind, flies tied so elegantly that neither fish nor fisherman can resist them. Some of us are firmly and permanently in love with all the surroundings of trout streams or lakes — bird and animal life, trees and bushes, reeds and water grasses, gravel and rock and sand, the water itself, and the way it moves and sounds.

All of us take with us to the waterside some blending of all these loves and fascinations. But there is nearly always something else, too, more subtle and more complicated. We are hoping to create or recreate some ideal situation. This may vary from time to time, changing in its intensity or emphasis. But I am convinced it holds its general shape throughout a trout fisherman's life.

I was raised on a south-of-England chalkstream, and by the time I was fourteen or fifteen my first ideal situation was firmly established in my mind. I sought it assiduously. A big fish (which meant anything over a pound) rising steadily, upstream of me and well away, probably close under a grassy bank or overhanging bushes. If I could cast a fly above him and have it float over smoothly and closely enough to bring a confident rise, the ideal was pretty well realized. If the fish was then firmly hooked and brought to bank, so much the better. But the ideal was in the scene and in the action up to the setting of the hook. What happened afterwards in those weed-choked streams was often anticlimactic and even a little untidy.

I had another ideal in time, dimly formed from the reading of many books about fishing: that of the downstream wet fly, sweeping across fast, broken water in a big stream, the heavy strike and instantaneous first run of a big fish. This was realized almost as soon as I came to the Pacific Northwest, best of all in the rapids and runs of the Nimpkish River on Vancouver Island. Here I found three- and four-pound searun cutthroats and rainbows feeding in the white water. They were not numerous, but a hard day of wading and casting would yield me five or six fine fish, and for the first time I knew trout that would run line well down into the backing from the strike and jump half a dozen times before they were beached.

I was usually alone on these expeditions and I knew, even while I was camping under the great spruce trees or in one of the bridge-crew shelters up on the logging track, that I was realizing another ideal that would always be important to me. These were undiscovered waters from a fly fisherman's point of view, their possibilities no more than vague rumours in nearby logging camps. I had to learn the water and the way about it, find the fish and their preferences for myself. An eighteen-year-old enthusiast could scarcely have asked for more than that great, gleaming river all to himself.

Big streams, heavy wading, fast water, and the wet fly became the standards of my fishing from then on. On the Nimpkish I had usually fished with two flies on my leader, one bright, one dark. I soon cut back to a single fly, but otherwise found little reason to change, even to catch fall-running coho salmon and winter steelhead. Much as I loved all these splendid things, though, I still carried my box of chalkstream dry flies wherever I went. Every so often I found myself recognizing something like the old ideal, the perfect dry-fly set-up. Whenever I did so I quickly profited by it, sometimes spectacularly. Once I saw sea-run cutthroats rising quietly along the shallow edge of a stream. Wading up outside them, and casting in toward the overhanging alders, I took six or eight perfect fish on a No. 16 Olive Dun in less than an hour. One July morning, at the edge of a fast run where a small stream came in at the head of a lake, I saw a big fish roll. He took a small English dry fly, as did three others, all big fish that had moved up from the deep lake to follow the sockeye salmon.

From chancing on such opportunities, and putting them to use, it was a short step to looking for them and even trying to create them. I learned about hair-wing dry flies and their floating qualities in fast and broken water. I searched riffles and runs and glides from downstream, with a fly dancing over the surface, instead of from upstream with a wet fly swimming under the surface. It was the old, original ideal, but grown much larger. Instead of the quiet interception of a fifteen- or sixteen-inch brown trout I was looking for the lazy, breathtaking roll of a ten-pound steelhead.

I am oversimplifying, of course. I have not abandoned wet-fly fishing, nor do I forget for a single moment the sharp pleasures of the sudden heavy pull and the first wild run away from the strike. When the conditions are right for this I take advantage of them with little hesitation. But the other thought is always in my mind, the moment when time stops utterly, that moment between the slow break of a powerful fish to the fly and the lifting of the rod to set the hook.

Chile and Argentina, the latter especially, have some of the finest wet-fly streams in the world, where big brown trout, rainbows, brook trout, and landlocked salmon, live and feed under ideal conditions. I have fished many of these streams, by no means neglecting the wet fly. One of my favourite Chilean streams is the blue-water Petrohue, flowing out of Todos Los Santos Lake. It is a big stream, wild and tumbling over rock ledges and among great boulders. One can wade it with some confidence, because the footing is generally good. There are places where every lengthening of the cast, every step farther into the fast water, seems to bring another good fish, rainbow or brown. Yet I remember best of all a little spring-fed tributary forcing out from under the Osorno volcano, where trout were rising in weedy water among grassy islands. The water was shallow and very clear and the fish, in spite of their steady feeding, were wild and shy. I fished floating flies upstream with care and caution and had fair success there until one evening when nothing went right. I realized belatedly that the fish were nymphing, bulging the water with their backs and tails but never breaking to surface-feed. In what was left of the evening I took three good fish on the upstream nymph and promised myself to do much better the next day. Fifteen years have passed. I never did get back there, but I still remember that little side stream as a trout fisherman's ideal.

The Argentine pampas streams are superb, clean and brilliant, leaping and gliding and driving their way among the brown hills as I imagine the streams of the Rocky Mountain States did long ago. The Chimehuin is not least among them. I have waded into its rapids and taken two- and three-pound browns and rainbows almost at will on sunk flies, delighted by the ease of it and the wildness of the fish. But I remember best of all a long, narrow island that split the river, forcing a deep dark flow smoothly under a cut-bank. It was dry-fly water and nothing else, but I was certain no artificial fly had ever floated over it before. It was midday, with a hot sun, and I saw no fish rise. Halfway up the glide, almost against the sharp break of the bank, the five-pound brown trout dimpled and took down a floating Purple Upright. Thirty or forty yards farther up a three-pound rainbow came to the same fly much less tidily, but he was just as securely hooked.

The magnificent Martinez Pool lies just below the junction of the Liucura and Trancura Rivers in Chile. It is enormous. It holds many fish and some very big ones. The two rivers burst in at the head with a rush that only gradually subsides through the quarter-mile length of the pool. The water is deep on both sides, so wading is difficult and limited, and I have never been there when it was not windy. It is logical to fish a wet fly there and I have done so with good effect. But the lower half of the pool, its relatively smooth surface marked here and there by great sunken boulders, tempts the dry-fly enthusiast in spite of its difficulties. Fish rise here and there, and a dozen different places promise smooth exciting floats — if they can be reached.

I have never managed to do things just the way I think they should be done in the Martinez Pool, though I have taken some nice fish there on floating flies. My sharpest memory of the pool is focused on a huge rock that just breaks the surface of the water near the tail of the pool, thirty or forty yards out from the north bank; a McKenzie Yellow Caddis sliding along it, floating high; a long, gleaming bar of silver, twisting slightly to catch the light, seen momentarily against the dark rock; the fly gone, my rod raised into solid resistance, another gleam of silver and the fly coming back to me. The rainbows are very bright in the Martinez Pool and this one was a full ten pounds.

So we search for trout in many ways and many places, each of us, I suspect, with some secret inward vision, subconscious as often as not, of what trout fishing really is. We will settle for less, often much less, and we may even find other, unexpected experiences more brilliant than the one we seek. I have many friends who never stop travelling in search of trout. They speak to me of Alaska and Africa, of Normandy, Czechoslovakia, Yugoslavia, of Scandinavia and Tierra del Fuego, of trout in Ceylon and Kashmir and Colorado, in New Zealand and New York State. They are all enthusiasts and, though I know some of their individual special enthusiasms, I would not try to guess the ideal that each one surely carries in his mind.

Frederick Halford believed he was not trout fishing unless he was casting to a rising fish the most perfect imitation he could create of the natural fly that the fish was feeding on. Few of us choose to limit pleasure so narrowly. Yet there is something of this spirit in all trout fishermen. We aim for the natural, unsuspecting response, which tells us that everything — choice of water, approach, cast, leader and fly — has been perfectly planned and executed. But if we are sometimes short of perfection and the results are good anyway, we accept them happily enough.

Last summer I fished for a few days in northern British Columbia, in wild waters far from roads, whose possibilities had been little tested. The weather was bad even in early August, with snowstorms in the mountains, mud slides and night frosts along the river. Yet I felt again all the dawn-fresh excitements of discovery that I had known forty years ago.

Here, in the calm outlet of a stormy lake, rainbow trout

and Arctic grayling were rising. The fish were bright and clean, fourteen to sixteen inches long for the most part, by no means suicidal, but easy enough to rise and hook. In the upper pool, as often as not, a five- or six-pound Dolly Varden came up to chase the hooked fish wildly through the shallows.

Later, down along a big river, storm-swollen and thick with silt, there was a sliver of clean, blue water from an entering creek. Grayling rose there, faithfully and accurately, to small flies that danced on the broken water. Grayling are not trout and they do not fight like trout. But they are as beautiful as any fish that swims and I often think trout could learn a good deal from them about surface feeding and honest response to a well-presented artificial.

At the head of the big river is a small lake, more than four thousand feet above sea level, the mountains around it seeming like low hills in spite of the summer snow on their slopes. A good stream enters at the head of the lake, straight from the glaciers and snowfields, across a flat cut up by tracks of woodland caribou, moose, wolves, and grizzlies. The stream was high and milky with storm water, but good-sized rainbows were jumping again and again where it entered the lake. I had only a little while to fish, so I used some of it to put up two rods, one with a sinking line. Then I offered the jumping fish a dry fly, as was only fair and reasonable. They jumped on, paying not the slightest attention to it, which did not surprise me. I offered them a credible wet fly; a sedge of some kind, still on the floating line. They were not interested. I turned to the other rod, which had a No. 6 Silver Doctor to go with the sinking line. Even so, I had to get it well down. But once I had the depth there were fish at every cast, strong, deep-bodied, splendidly clean rainbows of sixteen and eighteen inches that jumped like the wild things they were.

Later in the year I fished for a few short days in Wyoming and Montana. It was too early, I was told. Stream temperatures were too high and the big fish had not moved up from the reservoirs. I caught some smallish cutthroat-rainbow hybrids in a small but beautiful stream in Wyoming and a few more impressive fish from the big rivers of Montana. But fish were scattered and hard to find. Then, one late afternoon, I found myself on the bank of a spring-fed meadow stream, a tributary of the Madison, under a high, hot sun. Only the water was cool, and very clear, rippling across flats and around great banks of tight, green weed, gurgling along curving cutbanks, flat and smooth in the eddies. A few May flies were coming down and a much larger hatch of at least three different types of sedge was just starting. Trout were rising, quietly and efficiently, and though none looked large I knew there were big fish in the stream.

I went to work cautiously, keeping well down, throwing a No. 16 fly on a 5x leader. Three fish came short to it before I hooked a pretty little ten-inch brown. I was satisfied then that they were no fools and began trying to match the two smaller sedges that were hatching.

Nothing was easy, but when I did everything right I rose and hooked fish. Time passed swiftly and soon the sun was touching the mountaintops. I had released eight or ten fish, all browns, none over fourteen inches, most of them around ten inches. But the sedge hatch was on the water now and fish were rising everywhere. I missed two and hooked two in a roily piece of water below a small island. It was dusk by then and a good fish rose in backwater above the island. I made three perfect casts in succession to the spot, laying my leader over a weed bank. Nothing happened.

In a glide tight under the curve of the far bank I saw a tiny rise and put my fly a foot above it, brushing the grasses. The tiny rise came again and my fly disappeared. I struck and moved back from the bank fast. The fish ran upstream, taking line from the reel with a firmness and speed that commanded respect. Then he turned and came back, very fast, spreading heavy ripples over the smooth, dark surface. I recovered fast, but there was no fish there, only the fly.

As I dried off the fly there was another tiny rise, thirty or forty feet farther up against the grasses at the edge of the same glide. I covered it, the fish rose, I struck — and a four-inch brown trout came flying through the air toward me.

There was time to rise and lose one more good fish, then it was dark and time to go. It was only later I realized that there, in the rangelands of Montana, under the high mountains, I had fished an evening rise exactly like those of the gentle streams of my youth.

Of such things as these, and many others, are a trout fisherman's days and ways. It is a sport that can never grow old. We follow the traditions, but do not hesitate to bend and twist them to our needs. We dream dreams and make plans and nearly always fail in the execution of them. We surprise ourselves often, perhaps because we know so little about it all, perhaps because we are such simple souls. But wherever we go in the world we find other men speaking the same language, planning the same plans, dreaming the same dreams. And one of the big four — brownie or brookie, cutthroat or rainbow — is the cause of it all.

2
Watch the Creek Mouths
(1955)

Wonderful things can happen off the creek mouths. Sometimes big cutthroat trout swirl and slash the water everywhere, chasing needlefish or salmon fry; sometimes a ten- or fifteen-pound silver salmon quietly accepts a No. 12 fly intended for lesser game; sometimes the big cutthroats hunt deep where the river channel winds across the tide flats; sometimes they are up on the flats with the tide itself, ranging and searching; sometimes they dimple quietly as the flood makes, taking in floating flies brought down on the freshwater stream. And sometimes, inevitably, the fish are quiet or simply not there; but even then the creek mouths are wonderful, with the cry of killdeer and meadowlark and yellowlegs, the hunting of the osprey and eagle and otter, call of the loon, grate of heron, or whistle of pigeon. Man and beast and bird and fish all have a place and a part to play where fresh water runs into salt.

The creek mouths I am thinking of are along the Pacific Coast, perhaps especially along the coastline of Vancouver Island, though there are plenty to be found also on the mainland coast of British Columbia, along the Washington coastline, and presumably in Oregon and Alaska as well. They are the mouths of the lesser streams, branches and brooks and burns and rivulets, flowing directly into salt water. Some a stride's width, some jumping width, some a little wider than that. None looks fit to produce any trout much larger than six inches and most are so bushed and awkward farther upstream that to fish them anywhere but out on the tide flats is misery and frustration. Yet they are producing trout streams, often better producers in proportion to their size than much larger and more famous waters. And they sometimes show sport when the named and famous rivers seem barren and empty.

Generally speaking, Pacific Coast trout streams are not great growers of fish. They are spawning beds and nurseries from which migrant trout, both cutthroat and rainbows, go to sea to make their growth, returning only to spawn or at times of special abundance, when the stoneflies and salmon fry are hatching in early spring or when salmon are running and big sedges are plentiful in the fall. Streams that do not have true summer steelhead runs can be very quiet and almost empty of sizeable fish in the midsummer months. Some of us persist in searching then for the rare fish, steelhead or cutthroat, that chances in, but we do so more because we love the streams and they are close at hand than because we expect great fishing. Wiser fishermen turn to the lakes or to the salmon in salt water and so, usually, does the trout fisherman on a visit from some other part of the continent.

But such a visitor is often a puzzled and disappointed man. A keen fly fisher, he has heard of the sport that Pacific Coast trout can show and he knows the names of the famous streams. June and July are logical holiday months, and experience on other trout waters tells him he should be able to find at least a few fish around. Here is beautiful water, his own good skill, his tried and tested tackle. Yet pool after lovely pool yields nothing at all and he can only wonder whether to blame his methods, his gear, or simply his lack of local knowledge.

Local knowledge, back at the hotel or the auto camp, will

tell him of the salmon waiting to be caught outside or of the lakes upstream or perhaps of trout feeding well in the estuary of the famous river. It may or may not tell him about the creek mouths. Yet for a really keen fly fisherman there are few places more exciting or more rewarding.

In a sense the creeks are no better than the big rivers; it is all a matter of proportion. Here too the trout will be feeding in the estuary, perhaps even outside it. But there is less water to search through than in the estuaries of the larger streams; the trout are just as big, sometimes bigger; a man can work the place from his feet and keep track of what he is doing; and there is a fine chance of pleasant surprises — an early running silver or pink salmon, or a school of Dolly Vardens.

I have suggested that a creek need not be large. If it is running at all in midsummer, it may be worth a trial. Four or five years ago my friend Letcher Lambuth and I decided to fish a small stream that drains a fair-sized lake in one of the inlets on the B.C. Coast. When we reached the stream — we were in a small inboard boat — we found a logging camp established at its mouth and a maze of log booms blocking the channel. Neither of us felt like fishing there. I had noticed an alder flat with a small valley running back from it on the far side of the inlet, about three miles down, and I felt sure we should find a creek there. We ran back to look.

It was a creek all right, flowing out through the alders, across the sea grass flat, into a crescent bay behind two rocky islands. It wasn't necessary to step across it — one could have shuffled through it in low shoes without getting a foot wet. But that was a very dry season; the valley and the creek bed itself suggested a moderate flow in normal seasons, quite enough to support a run of spawning cutthroats and probably a few silvers and pinks as well. I was hopeful.

Between the rocky islands and the creek mouth the bay was very shallow, but a curving drop-off ran from the islands to each point of the bay. We ran the boat to the west side, where the drop-off was exposed to the flood tide. As we came to it, before we had put out a line, a reel started to scream. It was my reel. The No. 6 Silver Brown I was using had come loose from the reel seat and flipped overboard. A five-pound silver salmon danced away with it as Letcher and I struggled to untangle the rod, and ten minutes later I netted him — lucky, as Letcher did not fail to point out to me, not to have lost rod, reel, line, and all.

Through most of that sunny afternoon we drifted the boat back and forth along the drop-off between the islands and the point of the bay, watching the needlefish shine silver under us, casting our trout flies, never sure whether the next cast would yield a two-pound cutthroat or a five- or six-pound silver.

That was in the second week of July and it was unusual because the feeding silvers were with the trout, close around the creek mouth so early in the season; but it was typical of creek-mouth fishing exactly because something unexpected had happened, to set a new pattern of expectation for future times.

Tide is a factor in fishing the creek mouths, and it is a difficult one for a stranger. The ideal stage of tide can vary from creek to creek and with the season. On an unfamiliar creek it is a good general rule to pick a long run-out, fish the last two hours of the ebb, then fish or watch closely through the whole of the flood. The low tide gives one a good chance to see the shoals laid bare and the creek channel completely revealed; fish are usually active on the last two hours of the ebb and they are also confined within comparatively narrow limits at that time and on the start of the flood. Feeding activity may slow up as the tide makes, but there is a good chance that the fish will come on again at some stage of the flood; working a full range of tide is the quickest way of learning local peculiarities.

Though I have taken fish from tidal waters at every stage of both ebb and flood tides, I prefer the last hour or two of the ebb because it is the pleasantest time to fish as well as one of the most likely times. But there are very active feeding periods on half-tides at some creeks; and the approach to full flood, especially late in the evening, may be the best time of all. On big summer tides the fish work into shoal water at the edge of the flats, hunting feed in the flooded grasses, and often follow the salt water into the highest pools of the creek bed proper. A feeding flurry is possible at any stage of a rising tide.

The fish in the creek mouths are by no means easy to catch. Once in a while one finds a great concentration of them, rising and slashing all over the surface, and for a brief spell every cast brings a quick response; sometimes they are unseen, but feeding so faithfully that a deep fly, slowly recovered, brings a succession of solid strikes. But on most days they are scattered and capricious, hard to cover and hard to please; and on a few days they are abundant, obviously active and superbly contemptuous of every artificial.

Trout food in the creek mouths falls into three or four different groups. Usually some kind of forage fish is available — salmon fry in spring, needlefish and small herrings in summer, bullheads and other small scrap fish throughout the year. So silver-bodied flies are important. The most popular is made of green and white bear fur with a tail of scarlet wool, but many others are good. My own preference is for flies of my own like the Silver Brown and Silver Lady, but I have done well with Cummings' Fancy, Teal and Silver, Silver Doctor, and even an Alexandra. Yet another useful fly, highly recommended by John Atherton in his book *The Fly and the Fish*, is the Spruce Fly. Atherton gives a clear description of how the fly must be tied, with the two badger hackles of the wing set back to back, to open and close as it works through the water. It seems likely the fish take the fly over a bullhead.

The next important group of food organisms in brackish water are creatures like the sow-bug, *asellus*, and the sand-hopper, *gammarus*. For *asellus* I make a fly with a wool or seal's fur body of different colours, built up to nearly a quarter of an inch thick on a No. 8 hook, with peacock sword tied over the back and a badger hackle wound on the head. For *gammarus*, hen pheasant wing, olive seal's fur body, and blue jay hackle make a logical and effective combination; but here again the colour variations are innumerable, and I suspect flies like Mickey Finn and Yellow Peril are taken for sandhoppers.

The third group is plankton, or near it, and generally impossible to imitate, though the fish are sometimes full of it. In some creek mouths there are quantities of tiny, transparent shrimps, which hold position in the current and so are not really plankton. The trout feed on them and are very hard to take when doing so. I make a fly on a No. 8 or 10 hook with a short, silver body and a sparse wing of summer duck strips or barred monkey fur, and quite often persuade shrimping fish to take it; but the same fly takes salmon and large cutthroats that are not shrimping at all, so perhaps I should not give it too much credit as a shrimp imitation. It once hooked me a big yellow-striped rockfish on an early-morning ebb, then a coho jack, a cutthroat and, on the start of the flood, a bright female pink salmon that had just turned into the creek from the kelp beds.

Surface insects are probably the least important of the food groups so far as the fish are concerned, but they are important to the fisherman because creek cutthroats take floating artificials quite freely and faithfully whenever they are feeding on floating naturals. I have never found choice of pattern a very serious problem at such times. A brown and white bivisible on a No. 10 hook should be good enough. Many other patterns take equally well and I suspect this is because estuary trout are seldom feeding selectively or very seriously on the surface; they usually begin to rise on the flood tide, in tiny dimples and quite lazily, as though anything floating were a luxurious change from the more active chase of underwater creatures.

Any trout rod of around nine feet, weighing five or six ounces, is about right for the creek cutthroats; it should carry an HCH or HCF line. The reel should have a light check, and at least a hundred yards of backing — not for the cutthroats, which will rarely take more than twenty yards, but for the silvers, which can take most or all of the hundred. Leaders should be tapered to 2x or finer in bright weather with small flies; but it is usually sound to use 0x or even 9/5 with No. 4 or No. 6 flies.

Unless the fish are showing on top of the water, a fly sunk deeply and recovered slowly should do best. But in spite of this I nearly always use a well-greased line, for two reasons that seem to me quite good enough: a floating line is far easier and more comfortable to fish with, and it gives one a chance to see the following wave and swirl of a good fish taking — an especially keen thrill on the glassy surface and slight, dragging currents of low-water slack.

The creek mouths, like the rivers, are at their best in spring and late fall. But they produce or attract enough food to bring fish in at any time and to keep at least a few around all the time. That is enough. Good fishing, after all, is really no more than an honest chance at a few good fish. If the fish will rise or show themselves, so much the better; if they will take hold of fly or bait, so much better still. But even if they do nothing constructive for him, the wise fisherman can ply his rod with conviction and find a thousand pleasures in land and sky and water so long as he has reason to believe there is a chance they will.

One of my favourite creeks flows into tidewater past an old deserted homestead, among the rotting piles and broken wharves of an abandoned logging camp. Purple martins nest in the old piles; an osprey hunts the shallow bay, and often an eagle waits in some tall tree to swoop down and rob him of his fish; seals play out beyond the creek mouth and come in to stare at the fisherman wading long-legged where only herons should be; always there are shore birds, sandpipers, yellowlegs, snipe, dowitchers, curlews, and plovers, passing in migration or in casual summer flight.

I go there as often as possible, whether the fish are in or not; if I think nothing is likely to be around, but the day is pleasant and I'm free to go, I simply tell myself it's an ideal time to learn something new about the creek. And sometimes it is.

I was there one early morning late in July last year, expecting very little because fishing had been slow everywhere for several weeks. It was a lovely, still morning, the start of a hot, clear day, and the tide was far out, leaving the creek to wind in its channel all down the sandy flats to the eel-grass beds at the edge of the drop-off.

I worked a small shrimp fly on 2x gut rather carelessly down along the channel, searching for sign of a fish. One or two twelve-inchers followed and plucked at the fly, but that was all. At the end of the flat I waded out as far as I could and began to search more thoroughly, letting the fly carry the gut well down before recovering. As I waited on the third or fourth cast the floating line drew suddenly tight. I lifted the rod and a moment later a silvery fish jumped clear out, fell back, ran line, and jumped again. I judged him at seven or eight pounds, certainly a silver salmon, though it was much too early for silvers to be hanging around the creek. Ten or fifteen minutes later I beached him — six pounds, a feeding silver that had chanced across the creek's outflow as I set my fly there.

It was more than a good start to the morning. The quiet take to the unworked fly, fifteen minutes of gentle-handed anxiety with a six-pound fish running wild on 2x gut, are reward enough for any morning's fishing. But a two-pound cutthroat took my fly in the same place, and as soon as the tide made a little, I began to work slowly back up the channel. Nearly halfway up, by the sunken piles, a long V-ripple came fast at the fly and turned in a bulging swirl. The fish ran hard from the strike and jumped with thirty yards of backing out.

Golden this time, not silver, so I knew him for another cutthroat. He also came safely to the beach, an early-running harvest fish of well over three pounds. Fifteen or twenty minutes later I beached a twin, released her, and called it a morning. Nine herons along the edge of the flat were still undisturbed; two wigeon, far too tame, stood with heads tucked under wings not ten feet from where I took my rod down, and the hooded merganser brood swam in line ahead along the far side of the channel.

The creek mouths may be patchy and uncertain; they often are. But they have very special charms of their own. They are never the same for two consecutive days or two consecutive hours, because tide is always changing. And in each and every stage of tidal change there is movement of feed and fish, new water to search, new conditions to meet, hope revived for any fisherman. They may not be an all-sufficient substitute for the sweeping pools of a full flowing river, but they hold fish, big fish more often than not, and fish that set some nice problems for the fisherman. When Pacific Coast rivers are slow, check on the nearby creeks.

3
The Quinault River
(1958)

My wife calls the Quinault "that dark river." But she thinks of it as it is in November, with rain and mist rolling in from the Pacific Ocean, the heavy woods dripping, the current of the lower reaches creased with leaden lights and splashed with rain. In the late fall my wife and I used to run upriver with Herbert Kapulman from Taholah to visit a gill net at the site inherited by his wife, in the difficult eddy under a dark clay bank. As he pulled in lead line and cork line, shaking out leaves and drift, mending where necessary, picking out the big bright silver salmon and occasional early-running steelhead, Herbert would talk freely and well to us about his river and its promise of some kind of fish nearly every month of the year. "You should come fish," Herbert would urge me. "We have a pretty good river here."

The Quinault River rises on the slopes of the high Olympics, flows westward through Lake Quinault, and empties into the Pacific Ocean at the Indian village of Taholah, behind Cape Elizabeth. The upper river is made up of the North and East Forks, which join above the lake. The lower river flows thirty-five miles from Lake Quinault to the ocean, through the Quinault Indian Reserve. U.S. Highway 101 crosses the river near the lake's outlet, and from there to Taholah no public road goes near the river. Puget Sound and the roadless Olympic Mountains stand directly between Seattle and the Quinault. The Quinault is accessible, yet remote, and so it is one of the least known and least spoiled streams in the United States.

The fisherman has two approaches to the lower Quinault: to work up from Taholah at the mouth or work from Lake Quinault downriver. Working from Taholah has one advantage: the ocean beaches are right at hand; for an interval away from fishing, the beaches are an exciting pleasure. There are times in early spring and fall when working out of Taholah in the lower and middle reaches of the river may be the better plan.

Generally speaking, however, working from Lake Quinault down is probably the better choice, because the first twelve or fourteen miles downriver from the lake hold more broken and varied waters. The lower river is entirely bound by the Quinault Indian Reserve, and no one may fish it without an Indian guide. Since there is no way to get to the good water without a canoe, this point is as much a necessity as a law. In any case, the guides, their canoes, and the reserve itself, are a good part of the pleasure of fishing the Quinault. The Quinault Indians have always been fine canoemen. Like the Nootkas of the west coast of Vancouver Island, they were whale and sea otter and fur seal hunters who took their ocean canoes twenty or thirty miles offshore into the Pacific swells.

The Quinault guides are very casually organized and are very independent characters. Judged by the highest standards, they are canoemen rather than fishing guides; but they are fine canoemen and have had a wonderful record of safety over many years.

What does a trout fisherman look for on the Quinault River? Primarily cutthroat trout, running up from the sea or dropping back from the lake, for there are few, if any, resident trout over ten inches. After those, perhaps a summer steelhead. From August on, possibly jack salmon (small Chinooks or silvers) or black salmon (large Chinooks). And after the first fall rains bring the river up, silvers, early-running winter steelhead and harvest cutthroats.

The best month for trout fishing is July, when cutthroats are abundant and the river is in good shape. I suspect that the fall fishing after the first rains may be just as good if not better, especially if the silver salmon are taking freely. But few people fish at that time.

When I set out to fish the Quinault toward the middle of September last year, I had an open mind. I was hoping for the usual fall variety of a Pacific Northwest migratory stream — some big cutthroats, a few jack salmon, perhaps a steelhead or two, or an early silver; perhaps, beyond these, something quite new and strange. But I admit I wasn't too much concerned. I have loved rivers and canoes from childhood. I admire and enjoy the Northwest Indian people. I chose to try the more varied waters close to the lake. Accordingly, I looked up Jonah Cole, the sixty-seven-year-old dean of the guides who operate from Amanda Park on the lake. I found Jonah working on a new canoe, still shaping the outside with an axe, though it was almost ready for hand adze and plane.

"The river is very low," he said. "Lowest I've ever seen it."

"Too low for fish?" I asked.

Jonah shrugged. "Very low."

"There'll be some," I said.

"There'll be some," Jonah agreed, and I began putting up my rods. Another canoe waited at the landing, an eighteen-h.p. Evinrude on the stern. It was very beautiful — Nootka pattern, flat-bottomed with straight, flaring sides, vertical stern and high prow, over twenty-seven feet long and about forty-two inches wide, black and slender and graceful, yet strong and stable, perfectly adapted for river work.

I picked up Jonah's hand adze. It was D-handled, of the old pattern, with a steel blade in place of stone and reinforced at several places with friction tape. The head of a man with a sea otter headdress was carved on it, which I thought unusual. The Quinaults were not great ornamental carvers.

"Belonged to my grandfather," Jonah said. "Must be a hundred years old. He was a Hoh River man. I am a Hoh River man, married a Quinault woman. Not many Quinaults left."

I asked him if he knew that his forebears on the Hoh River had wiped out a boat's crew from Quadra's little Spanish ship *Sonora* in 1775, and another a few years later from Captain Barkley's fur-trading *Imperial Eagle*. Both boats had turned into the river for fresh water.

"Killed all the white men?" said Jonah. "Ho!" He was amused and by no means displeased.

"I guess they just didn't understand each other," I added. "But the Hoh River people were tough men. They had hair on their hearts."

"Hoh River men all over the country now," Jonah said proudly.

That evening Jonah and I went down the river three or four miles, to the head of Sherman's Pool. Young steelhead from six to ten inches were everywhere, some of them silvering up for the journey to the sea. They rose freely to sunken or floating flies, though it was possible to avoid hooking most of them. One fine fat cutthroat of about fourteen inches came to a Silver Brown, fished wet. Jonah, casting a big Double Tacoma with red yarn and a piece of fresh mussel on the hook behind it, caught two others of about the same size. Using an eight-foot steel rod and a Shakespeare bait-casting reel, Jonah soon showed me he was an artist. Short or long casts went out with beautiful accuracy, under limbs, tight up against logs or brush, into every likely place. All three fish we hooked came from deep water. I was pretty sure we were going to have to hunt and hunt hard for them.

Running up the river in the fading light, I didn't really care. The long canoe performed beautifully with the motor, riding through the long, still reaches with scarcely a disturbing ripple at twenty-five miles an hour. In the rapids Jonah was a master, slowing the motor, using it exactly as a pole to ease the canoe against the current and straighten it into position, then opening up the throttle for the big push over the lip, easing off again to keep the propeller from hitting on the shallow of the break.

Through the next three days Jonah and I conspired together against the fish and worked together against them as best we knew how. We ran down nine or ten miles to Eagle's Lake Pool, then twelve or fourteen miles to Burnt Hill Pool and below.

We caught some fish, perfectly conditioned cutthroats, apparently fresh from the sea, from twelve to sixteen inches, always in the glides and deeper places, never in the riffles and runs where they should have been. At Sea Lion City a big cutthroat came to my Silver Brown over quite shallow water and threw the fly almost at once. In Eagle's Lake, below the standing rock, a bright steelhead rolled to the same fly but would not come again. Burnt Hill Pool was full of jack salmon, though I had to descend to a thumbnail spinner ahead of an orange fly to get one. Jonah hooked two and broke in both, then suddenly pointed downstream.

One of the fish Jonah had hooked, rusty-red and ten or twelve pounds, kept rolling in midstream, but wouldn't touch my fly, even with the spinner. I looked up at the timbered slope of Burnt Hill across the river and thought of it as it used to be, burned clean and growing nothing but ferns. I thought of the young Quinault boys who used to come here at the end of their novitiate to keep vigil for the guardian spirit who would determine their lifework — the spirit who would give power to become a great hunter, medicine man, fisherman, a great whaler or hunter of seals.

The Quinault is a good-sized river, by which I mean that a good fly fisherman, at normal water, would be attempting to cover only about half the stream. Immediately below the highway bridge it is fairly well broken up by rapids and boulders, but from there on, as Jonah says, "it begins to flat out." There are innumerable runs and riffles and minor rapids, some of them quite swift and difficult, but there are also long quiet reaches with very little current and sometimes quite deep water, and the rapids and runs are gravel-bar breaks rather than rock and boulder breaks. There is little solid rock anywhere along the river.

This all adds up to a wonderfully pleasant and easy river to fish, with an almost infinite variety of water. There are few casting problems so long as one holds a high back cast to clear the gravel bars. Wading is safe and comfortable at most places, though felt soles and wading staff are desirable. Distance is nowhere a real problem; seventy feet will reach almost anything from any given stand and the canoe solves the difficulty of crossing to the other side of a straight run or reaching a glide under the far bank of a quiet stretch.

I put up two rods the first day and used them with complete satisfaction throughout the trip. The first is eight feet nine inches, heat-treated cane, weighing five ounces, and I used it entirely for wet fly, throwing an HCF Dacron line with modified forward taper, about forty-five feet from point to running line. The second is an eight-foot rod, weighing four ounces, also of heat-treated cane, and I used it entirely for dry fly, with a well-greased IEH Dacron line of the same modified taper or an HCH nylon floater, which I don't care for. I had plenty of backing on all reels, as I always do in Pacific Coast migratory streams. The cutthroats may not call for it, though a fresh-run eighteen- to twenty-inch fish probably will, but the possibility of steelhead, silvers and even Chinooks is much too real to be disregarded.

Because the weather was so bright, I mounted nine-foot, 2x (.009) leaders on both rods and fished that most of the time, changing to 0x (.011) with larger wet flies. I carried a wading staff in the canoe, but did not bother to use it — though it would have been handy in the Boulders and useful at the Burnt Hill log jam. In higher water I think I should have been glad of it in several places. A net is pretty well essential, but only when fishing from the canoe or in the Boulders. At most other places where one wades there are good gravel bars for beaching fish.

Here it is necessary to admit that the Quinault is a river without any real fly fishing tradition. Try as I would, I found no local fly fisherman who works the river regularly. Jonah mentioned several of his customers who use fly exclusively, or most of the time, but admitted that the majority are spinners or bait casters and a good many are not fishermen at all.

Because of this, there seems to be no such thing as a local Quinault fly pattern, and the local stores carry only a few standard flies. Jonah, instead of producing some local mystery from deep in his tackle box, simply asked: "Have you tried Royal Coachman? Parmachenee Belle? Silver Doctor?" Plainly the eastern tradition, unmodified by Coast conditions, except for one concession: "You tried bucktail?"

Another Quinault fisherman, Harry Bergman, has told me of success with a Carey Special, and this surprises me because it is primarily a lake fly, and I have not done well with it in rivers. But it is probably significant because the Quinault is full of crayfish, which the trout use, and the fly looks a good deal like a small one. My own most successful fly was the Silver Brown, which I first made many years ago especially for fall cutthroats and steelhead. The river has plenty of sedges and May flies and probably stoneflies in season; the sedges I saw varied from a small black to a large cinnamon. There are also plenty of salmon and steelhead fry and fingerlings and a fair number of bullheads.

This suggests that most of the standard Coast patterns, wet or dry, should do well enough. Such well-known flies as Skykomish Sunrise, Mickey Finn, Brad's Brat, and Lady Godiva are good general patterns. When fry are around — which is most of the time — Silver Brown, Silver Lady, and

Silver Doctor are good standbys. Steelhead flies, such as Harger's Orange, Queen Bess, and Golden Girl, often take cutthroat and jack salmon well. Add Dark and Light Caddis patterns, Western Bee, Spruce Fly, and Squirreltail — this last, with Carey Special, as a crayfish imitation — and the selection of wet flies should be adequate for any conditions. Hook sizes No. 4 to 10 or 12.

For both steelhead and cutthroat I prefer dry flies of the Wulff type in rather large sizes — usually No. 8 or 10, sometimes as large as 6 — but with darker dressings than for eastern use. McKenzie River patterns such as Beetle Bug, Light Caddis, Orange Caddis and McKenzie Special are also good, and the bushy Columbia River caddis patterns sometimes stir up an unwilling fish from deep water. Don't hesitate to drag or skip a dry fly, under control, on Coast streams.

The lack of a fly tradition on the Quinault is a little disappointing, but I think it has two main causes — the inaccessibility of the river except with a guide, and the extensive stretches of quiet water which make ideal lying places for trout and steelhead through the middle of the day and are far more easily searched with bait or lure than with fly. But there is abundance of magnificent fly water and there is not the slightest doubt that the stream often does produce really well for fly fishermen.

Starting at the highway bridge, there is a reach of deep, streamy water among big rocks that holds fish well at times. Below that the Boulders makes a series of wonderful resting and feeding pockets, every one of which may hold a good fish. Where the rapid levels off there are fine fast runs on both sides. The head of Sherman's Pool is nice water, levelling into China Flat, which is long and quiet, with two nice riffles at the lower end.

Beyond these are the Blue Banks, high, steep bluffs of blue clay and gravel, where Jonah once found two mastodon tusks after a slide had come down. There is deep water here, and the river circles almost completely back on itself. Lower Clay

Banks is a fine pool just above the logging railroad bridge, and Shorty's Riffle, another good one, is just below it. Jonah tells of taking ten winter steelhead in one day from this pool.

From there the river flats again to Prairie Creek (a nice run at the mouth) and on to Sea Lion City, which heads in a fine swift run under the left bank and spreads to a fascinating complex of gravel bars. Below that is Fred's Lake Pool (very deep and excellent holding water), then Eagle's Lake Pool, one of the prettiest reaches of fly water I have ever seen. It starts in a fan of swift runs over a gravel bar, each with a clear, sharp drop-off; the water collects from these to make a hard run on two sunken rocks and sweeps on to break against a high-standing rock below them; below the rock is a smooth, strong glide of deep water, gradually easing over until it collects in an easy run under the far bank; there are rocks and more good fly water below the tail of the pool, at least six hundred yards in all. It is a fly fisherman's dream of a pool that offers every classic situation of search except overhanging trees and brush. It fishes easily and, fish or no fish, makes one feel like a master because the fly is always doing what it should.

Another stretch of quiet water, and then you come to an excellent run and pool on the bend at Noisy Creek. More perfect water at Tsoluf Tsoluf, one of the places where the river narrows to a few yards, and the splendid holding water at Burnt Hill jam, then the fine pool below Burnt Hill itself. From Burnt Hill there are another ten miles or more of varied water before you come to the lower reaches above Taholah. It would take a lifetime to know it all and several years to try out all the good fly water.

I shall go back to the Quinault — not once, but many times, I hope. I am curious about the river and its fish. It seems quite clear that July is the best month, which is unusual in a Coast stream. Apparently there is a good run of fish in from the sea at that time and probably fair numbers of fish drop back from the lake as well. It must be a feeding movement, and presumably reflects some special activity of crayfish, salmon fry or insects, perhaps of all three. Jonah says the fishing holds on into August, then slowly drops off. In most streams the Coast cutthroat run begins to show in late July and early August, and builds from there.

The sea-run cutthroats, in any river I know, spawn sometime between November and March, with a peak usually in January or February. Milt and ovaries of the Quinault fish I caught suggested no variation from this pattern. If they leave the river in July, they must run back in again before Christmas to spawn. My guess is that they come up on the first real rise of the river, in late September or early October. By that time a few bright silver salmon should be working up into the stream with them. So I would happily gamble on a few days early in October.

But it is a splendid and challenging river at any time, and a wonderful place to be. In four days up and down the river, Jonah and I saw no other living man. We saw mink swimming and diving like muskrats, a family of three coons playing on a sandy beach until the old lady came and growled at us and chased them away, a beaver swimming quietly in the eddy of Sherman's Pool at dusk. Mergansers and kingfishers, grebes and goldeneyes, ospreys and flickers were along the river. Great, moss-hung spruce and poplar and hemlock trees stood along the banks; vine maple and crab apple were turning scarlet in the swamps and at the bends in the old river channels. The world was ours, to watch and explore, with the conviction that few men pass that way to disturb it.

4
The Splendour of the Run
(1968)

The Atlantic salmon is an aristocrat, the peerless fish of legend and ancient anglers' tales. I was raised to admire and respect him and some of my earliest and most vivid angling memories enshrine him. The first book I ever wrote was about him.

The Pacific salmons — there are six distinct species in all — are storied rather than legendary, one of the existing wonders of the world, massive in numbers, of immense commercial value, comparative newcomers in the angler's world. Few fish have been more studied or are better known to science; yet their comings and goings remain mysterious and often unpredictable, and with every new discovery the precision of their performance seems the more remarkable. After forty years of living beside rivers where Pacific salmon run, my curiosity about them is still unsatisfied, my respect for them has steadily increased, and my affection for them is unbounded.

That there should be six species, six distinct variations on the splendid theme of the salmon's life story, is in itself a biological wonder. No one doubts that the six species had a common ancestor; many believe that at some distant time before the ice closed off the Northwest Passage between Atlantic and Pacific, perhaps two million years ago, a common forebear of both Atlantic and Pacific salmons ran and spawned in the rivers of both oceans. In some way not yet fully understood the Pacific salmons have penetrated practically every river system tributary to the whole vast semicircle of the North Pacific Ocean. In most watersheds, large or small, several species are firmly established, each taking advantage of the characteristics of gravel and stream flow for which it is best adapted, and so completing an almost ideal pattern of use.

I have mentioned six species. One of these, the Japanese cherry salmon (*Oncorhynchus masou*) is limited in distribution to the Japanese islands and the nearby Asian coast. The cherry salmon is an interesting fish and I wish I knew more about him at first hand. He is said to be much like the coho in general habits, though more inclined to abandon saltwater migration and become a freshwater resident. He is a favourite of Japanese anglers and provides a great deal of sport. While I am not generally enthusiastic about the introduction of exotics, I have long believed that he might fill a useful gap in some North American watersheds; but so far as I know, no introductions have been attempted.

The other five species are all native to the Pacific Coast of North America and also to the rivers of the Asian side, especially those of the Kamchatka Peninsula, where they were first described by the great German naturalist, Steller, and the student Krascheninikov, in 1731. These five are the chinook, also known as the king or tyee; the coho or silver salmon; the sockeye or red salmon; the chum or dog salmon; and the pink or humpback salmon. All have different habits and life histories and differ in size and appearance, especially under spawning conditions.

It is not my purpose here to describe the differences between the species in any detail, but some appreciation of the variations is essential to the story and feeling of the runs. The pink salmon is the smallest of the five and has the simplest life history. It hatches from the river gravels in early spring as an inch-long silvery fry, goes straight to sea, and returns as a two-year-old spawning adult. In salt water it is a graceful, silvery fish of three to five pounds, fine-scaled, the forked tail heavily marked with oval black spots. Most streams have heavy runs in alternate years, with very light runs between. In the big years pinks may contribute a good deal to the sports fishermen's catch. They fight quite well, though they rarely jump, and I have seen individuals as large as eleven pounds. In fresh water, pinks lose their brightness very quickly and develop a spectacular hump and hooked jaws.

The chum, like the pink, goes almost immediately to sea and returns after three or four years at an average weight of eight pounds; a fish of thirty-six pounds was recorded at Petersburg, Alaska, in 1958. Chums run rather late in the year and are rarely taken by sports fishermen. I remember them as marvellously abundant thirty or forty years ago, but heavy commercial fishing cut down the runs, and the chums are re-establishing very slowly.

The sockeye is the most valuable commercial species. Red-fleshed, handsomely silver-blue, and elegantly shaped in salt water, the males become scarlet or crimson with olive-green heads as spawning approaches. Sockeyes average five or six pounds, but may weigh as much as ten or twelve. All major sockeye runs use river systems with large lakes, where the young fish grow through one full summer and winter before migrating to sea in the spring of their second year as three- or four-inch smolts. In Oregon, Washington, and British Columbia most sockeye return as four-year-olds. In British Columbia's Nass River and most Alaskan streams five-year-olds predominate. Sockeyes feed primarily on crustaceans in both fresh and salt water, and are rather rarely taken by anglers.

The two remaining species, the chinook and the coho, are the main concern of the sports fisherman. Chinooks run to the larger streams and spawn in flows roughly proportionate to their great size. The fry usually spend about three months in the river before migrating to salt water and the adults return after three, four, or five years of sea feeding; northern fish, especially those of the Yukon system, are known to return as seven- and eight-year-olds. The long, unbroken spell of ocean feeding makes for great size, and the returning four-year-olds usually average about twenty pounds. In streams which have a preponderance of five-year-olds the average weight is thirty-five to forty pounds. Fifty-pounders are not at all uncommon and sixty-pounders are taken every year. The world's record rod-caught chinook is a fish of ninety-two pounds, taken in the Skeena River near Terrace, British Columbia, in July 1959. But commercially caught chinooks of well over one hundred pounds have been recorded, including one of 125 pounds, so sportsmen can still dream of greater things.

Coho salmon run to even the smallest creeks and work their way up into the very headwaters. The young fish spend a summer and a winter in the stream, then two summers and a winter in the salt water, returning to spawn usually as three-year-olds. In this short time they grow to an average weight of eight or ten pounds. Any coho of twenty pounds or over is a very good fish.

This, then, is the broad material that makes the mosaic of the Pacific Coast salmon runs. But the pattern is infinitely enriched by the varied timing of the returns and by the specialization of individual races of fish within each species. One run of Fraser River sockeyes finds its lake seven hundred miles from the sea, another three hundred miles, another less than one hundred miles; and there are separate runs to lakes in between these distances. Many streams have chinook salmon runs in both spring and fall. The Columbia has, or had, three runs: the first entering in January, February, and March; the main run in May, June, and early July; the fall run in late July, August, September, and early October. Cohos enter the rivers in September, October, and November, but I know of coho runs that come in as early as June and July, to be followed by normal fall runs in the same streams.

The salmon are the life and character of all Pacific Northwest coastal streams, from the ocean clear back to the Rocky Mountains, unless some obstruction bars the way. February is one of the quietest times. The river bottom is flood-swept and clean. Last fall's spawning multitudes are dead and washed away, except for chinook vertebrae lodged here and there in the lee of boulders and perhaps a spawned-out female coho still holding onto life somehow. Winter steelhead are silver-grey shadows, hard to find and hard to see. Down among the sheltering rocks there are coho yearlings, hidden and half-dormant. Deeper yet, still under the gravel, is the new season's hatch of chinooks, cohos, pinks, and chums.

Out on the salt water a few year-round anglers troll spoons or plugs or herring for winter-feeding chinooks and two-year-old coho "grilse," now twelve to fourteen inches long. It is a quiet time here, though the winter chinooks are among the finest prizes — green fish, as the fisheries men say, far from their spawning and full of fight.

By early April the river is coming to life. Pink salmon fry are massed in the eddies or passing downstream, where the sea-run cutthroats and Dolly Varden char come in to meet them. Mergansers and goldeneyes hunt the river edges and bullheads crouch and peer in search of emerging alevins. Coho yearlings and young steelheads have moved up into midwater and chinook fry are spread out along the gentler flow at the edges of the runs.

By mid-June the young chinooks are two or more inches long, silvering over their parr marks and river colouration, ready to go to sea as ninety-day smolts that will provide most of the return in future spawning runs. Already the spring-running adult chinooks are in the river, hiding their thick, powerful bodies where the white water of the rapids breaks over into the heads of the pools. In the salt water the cohos are splendid little fish of three pounds or more, hunting and feeding to make the rapid growth of their final summer. Far out in the ocean, the pinks, chums, sockeyes, and chinooks are starting the long journey toward their home streams.

Well before the end of July, mature pinks and sockeyes have come into the river. Early in August the first fall chinooks move in, huge among the little pink salmon, nervous in the exposure of sunlight and shallows. In the salt water, gill-netters, purse-seiners, and commercial trollers reap their rich harvest while thousands upon thousands of small-boat anglers search along the kelp beds, along the drop-offs, through tidal eddies, or off the river mouths. Through September the runs build in the rivers to the spawning climax of October and November.

Saltwater sport fishing for Pacific salmon has become a very big business indeed in all coastal waters that are sheltered or partly sheltered and within accessible distance of roads. Private boats and rental boats and even party boats are out by hundreds wherever and whenever conditions are reasonably favourable throughout a season that extends at its liveliest from June until October. Hundreds of resorts and trailer camps are filled year after year with eager fishermen and their families. The catch is measured by hundreds of thousands of fish. Small children boat monster chinooks of fifty, sixty, or even seventy pounds and are duly celebrated. Salmon derbies are run off with frantic publicity and spectacular prizes. Freezers are filled and home canning machines work overtime. The overall impression is more nearly that of a gigantic summer carnival than the gentle pursuit of angling. Yet the waters are big, the fish are to be found in many places; mountains stand tall over the timbered islands and coves and channels; dawn and sunrise are breathless over the water; sunset and afterglow lengthen lazy evenings while the rod top beats and the slow line cuts the water.

The most popular of all today's rigging, I suppose, is the large flasher or attractor — up to twelve inches long by three inches wide — set several feet ahead of fresh or frozen herring, strip, cut, or whole. Sometimes a small spoon or plug, or even a bear-hair fly, replaces the herring. And there is a chunk of lead to match the angler's estimate of the depth at which he will find his feeding fish. This is cumbersome stuff indeed, but it is effective. It has developed, no doubt, because fish are less abundant than they were thirty or forty, or even twenty years ago. But fortunately salmon can still be caught by more acceptable means. Knowledge of tidal conditions and feeding times and places still pays off, and many a skilful fisherman can work a deep herring effectively enough to make flasher trolling seem slow stuff. There are times, too, when a small spoon or plug with an ounce or two of lead will take fish as fast as anything else, and even times when the trolled fly will take surface-feeding cohos as well as it did thirty years ago.

The coho is, by any standards, an outstanding game fish. He hits hard, often right at the surface, runs out a hundred yards or more of line like a flash, and jumps with splendid abandon on reasonably light gear. He rarely runs deep or sulks. When he is feeding near the surface on herring or sand-launce, as is quite often the case, he offers fishing chances at least as good as in lake fishing for really large trout.

Under these circumstances fly fishing can be effective, occasionally very effective. But in general, large streamer patterns are more effective and even these are better trolled than cast, because the fish want fast movement and will often follow for a considerable distance before taking. The livelier movement of a small lure or spinner brings a much quicker strike and chiefly for this reason light spinning gear is really the ideal way of casting for surface-feeding cohos. The cast is made quickly and easily, the retrieve is readily controlled and, though the strike may be less exciting and satisfying, the hooked fish is just as free to perform at his violent best.

The fly fisherman has another and better opportunity later in the season as the fish move into the estuaries of the small streams. Here he is on familiar ground. Sea-run cutthroat trout are moving in at the same time and he need scarcely vary his tackle or his methods to take both fresh-run pinks and cohos. The uncertainty of what to expect next is part of the charm, but there are times, too, when a newly arrived school reveals its presence plainly and an accurate cast brings an immediate boil and a wild, leaping run.

Chinooks in salt water are usually deeper and harder to find than cohos, and the overwhelming majority are caught by trolling or mooching rather than by more active methods. But the chinook is an impressive fish to hook and an impressive fish to bring home; few can resist trying for him. Even a twenty-pounder can give an angler the impression of being hooked into a runaway steer, and fish of thirty pounds and up have such formidable and enduring power that a light-tackle fisherman may wonder if he really belongs there at all. The first run is usually an uncontrolled, line-drowning surge of power, often semi-circular around the boat or intermittent, or both; but the later stages, with a half-tired fish boring and lunging, head down, tail up, can be even more testing.

Most of the really large chinooks are taken off the mouths of streams that have runs of unusually big fish, such as the Nimpkish, the Campbell, the Gold, the Puntledge, the Somass, all on Vancouver Island, or the Owikeno in Rivers Inlet, British Columbia. At this stage they are feeding little, if at all, and are very unwilling takers. But they are concentrated in fairly shallow water and at certain stages of tide and light can be persuaded to strike at a slow-moving plug or spoon. Faint-hearted stream-fishers may lack the hardihood it takes to sit through the patient hours that bring a strike, but the enthusiast clings to the thought that the very next strike may mean the seventy-pounder that will crown his career.

The most exciting sport I have had with big chinooks has been in the tidal pools of the big-fish rivers, where they can sometimes be taken by casting a large spoon and working it slowly near bottom. A forty- or fifty-pounder hooked in this way, still full of life and energy, is an altogether different proposition from one hooked from a boat in the salt water.

There is a real sense of achievement in stirring the fish to strike, and the river makes problems in handling a hooked fish that are far more intricate and demanding than those of open water.

Both chinooks and cohos are frequently taken in fresh water and in some rivers a big run of pink salmon can become a major nuisance to anglers. The chief reason why Pacific salmon are not more readily sought after in fresh water is that they are usually close to spawning when they come in and so rather quickly lose condition and colouration. Exceptions to this are the spring-running chinooks, which provide magnificent sport in a few streams, and the early-running cohos. Nearly all larger streams attract a few cohos that run in as much as two months before the main body of the run. These are bright, clean fish and often take a fly or lure, the latter especially, quite readily. I have even taken one ten-pounder on a floating fly and have heard since of other fishermen who have done the same.

A few Pacific salmon populations are residual or landlocked — that is to say, permanent freshwater residents. They remain anadromous, making their main growth in lakes and running into the tributary streams to spawn. In Pacific-slope waters the sockeye is the commonest freshwater resident and has been given subspecific recognition by some taxonomists. They are known to anglers as Kokanees, or "little-redfish," and in most lakes do not grow larger than twelve or fifteen inches, though retaining most of the characteristics of ocean-running sockeyes, including the red flesh and brilliant spawning colours.

Because of their immense commercial value, the Pacific salmons have been somewhat better treated than their Atlantic cousins. Yet the runs have been damaged to an almost incredible degree by the profligacy and stupidity of North American resource development. The earliest evil was probably gold mining, which destroyed thousands of acres of spawning gravels worth far more than the gold they gave up. Overfishing played a part on occasion. Irrigation diverted millions of fry and fingerlings into ditches and fields. Small dams blocked off thousands of miles of productive spawning water — always unnecessarily and at small profit to the people. Then there have been the big, insurmountable main stem dams. Probably the most destructive single factor, and a continuing one, has been the damage to streams and watersheds by ill-planned, or unplanned, logging. Finally, there is pollution — industrial, municipal, agricultural, silvicultural — continuing and growing in spite of the pious words of politicians, the double-talk of industrialists, the face-saving of pollution control authorities, and the pleadings of the few members of the public who bother to understand and care about it. Pollution can still mean the end of the salmon runs, just as it can mean the end of so many other good things.

Without discounting the dark possibilities, I am optimistic about the future of the runs. Overfishing is a thing of the past, even though there is still too much commercial gear in the water and some system of licence limitation is long overdue. Salmon management, which is essentially the balancing of catch against spawning escapement throughout the period of any given run, has become sensitive and sophisticated. It is backed by efficient enforcement and increasingly accurate forecasting. Understanding of salmon movements and salmon tolerances, even in the ocean, is steadily improving and with every improvement hope is brighter.

It appears that ocean disasters of one sort or another do affect the salmon runs and it is unlikely that the precise nature of these disasters will be understood in the very near future, still less likely that means will be found of controlling them. But the resilience of the runs has survived such disasters from time immemorial. So long as a few fish return to spawn and the fresh waters they return to are in good condition, the numbers of the run will be restored within a few cycles. It is in the fresh waters that permanent damage is brought about, and it is here that man can do the most to restore and compensate.

In a broad oversimplification, it can be said that there are two chief methods of attempting to restore or improve salmon runs. The first involves hatcheries, selective breeding, and artificial rearing areas; its most logical application is on watersheds, such as the Columbia, where good natural conditions no longer exist or have been heavily reduced. The second method is by working with natural conditions: it seeks to control stream flow and thus eliminate destructive floods and droughts and provide good rearing conditions; to control temperature; and to increase spawning areas by constructing artificial spawning channels with ideal conditions of gravel, flow, and temperature. These last measures can yield egg-to-fry survival rates from natural spawning almost as high as those of hatcheries, but the young fish produced are stronger and healthier, with generally higher survival rates at later stages.

The first of these methods is producing remarkable but uneven results in Oregon and Washington. Fish have returned in spectacular numbers, though sometimes passing through so rapidly to their point of origin that they have offered little opportunity to either commercial or sports fishermen. This tendency probably can be corrected. A far more serious defect is that hatchery selection destroys the splendid variety of stocks that have been produced by ten thousand years of natural selection and so leaves the future very much in doubt. I have heard on good authority that some hatchery stocks are no longer capable of successful natural spawning.

While artificial spawning channels also tend to select, they do so far less drastically and the factors that select, such as temperature and gravel quality, can be controlled.

In most of British Columbia and Alaska, and in all streams not too much damaged by silting and pollution or reduced by obstructions, I believe this second method holds tremendous promise. Flood control is possible on many productive streams at reasonable cost, and this alone can yield greatly increased and far more regular returns. Once control is established it is fairly simple to extend natural spawning areas by building artificial channels paralleling the natural river bed. The river bed itself can always be substantially improved to provide better rearing areas for cohos and chinooks and other stream-type species. One small watershed on Vancouver Island, the Big Qualicum, is already under full control for salmon purposes and is proving itself by substantially increased returns.

Work of this type, extended to cover the streams that are the principal producers of the cohos and chinooks on which the sport fishing depends, should not only maintain the runs but should in time increase them far beyond their greatest natural abundance. The rest is a matter of keeping out pollution, protecting headwater timber, and restraining the dam-building mania.

The glory of the Pacific salmon runs is in their superb adaptability and complexity, in the infinity of combinations and permutations they have evolved to effect their survival in such remarkable abundance. That they have also developed two or three outstanding species of game fish is a fortunate chance. Even as commercial fish, they are one of the world's wonders, and my keenest pleasure now is in watching them; in watching the pleasures and satisfactions, even riches, they provide for so many people; watching the fluctuations of their fortunes in the continuing struggle to survive. Above all, in watching the fish themselves, the runs that go out and the runs that return, and trying to understand a little more clearly the meanings they have for man.

5
Diplomat's Fish
(1958)

One early October evening, thirty years ago, I was standing at the mouth of a small creek on northern Vancouver Island with my friend and partner, Ed Lansdowne. I was nineteen then and had been in Canada less than a year; Ed was sixteen and Canadian-born; but we were of one mind about hunting and fishing, and at that moment we were both absorbed in admiration of a great hook-nosed coho salmon. He had come from the little pool below the falls, where the creek started its run across the tide flat to Johnstone Straits, only a moment or two before. His inducement, and this was the important thing, had been a full-dressed Silver Wilkinson Atlantic salmon fly. It was the first coho we had ever caught on a fly; we had always been told, and almost believed, that Pacific salmon wouldn't have anything to do with a fly.

"Well, you did it," Ed said as the spell of silent delight wore off. "They should have sent you out here instead of the other Englishman that gave the country away."

"What do you mean?"

"You know, Lord Aberdeen. The guy that gave Washington and Oregon to the Americans because the salmon wouldn't take a fly."

That was the first time I had heard the story that is still a favourite along the Pacific Coast, on both sides of the border. It didn't sound like very good history to me and I argued with Ed about it most of the way home. There we appealed to Ed's older brother, Buster. "Sure, it's right," Buster said. "The Oregon boundary dispute. Must be right. They told us in school."

There the story rested, while I went on fishing and caught scores more cohos on flies of many kinds, and saw them caught by hundreds on the bucktail and polar-bear flies that became popular in the early thirties. Every so often I heard the story again, never too well authenticated, often told as a joke without any real foundation at all. It remained a good story and somehow added sharply to the pleasure of every new experience in fly fishing for Pacific salmon. I remember a wet and windy fall afternoon on the Campbell, the river high and a little coloured, and the fish running so straight in from the sea that needlefish were still undigested in their stomachs. Coho after coho took a No. 2 Mallard and Silver and tore off downstream till my fingers burned on the backing. In the end I broke the barb off the hook and encouraged them to shake it after a run and a jump or two, so that I could feel the fierce pull and the first hard run again and again. And all the time I kept hoping that old Lord Aberdeen was looking down from heaven or up from hell to see what went on.

But the truth is always around somewhere. It had been sitting on my own library shelves, in Judge Howay's *History of British Columbia*, for ten or a dozen years before I found it. The learned judge tells the old story in a couple of sentences and dismisses it sharply, then goes on to quote from a manuscript of Roderick Finlayson, the commander of Fort Victoria, which is still in the British Columbia provincial archives. Finlayson is describing a visit of Captain Gordon, in command of *H.M.S. America* and "brother of the Earl of Aberdeen, then Prime Minister of England."

"Capt. Gordon," Finlayson wrote, "was a great deerstalker. We met a band of deer and had a chase after them on horseback. The deer ran for a thicket into which the horses with their riders could not penetrate and of course no deer were had.

"The Captain felt very disappointed and was anything but happy. I said to him I was very sorry we had missed the deer etc., and also remarked how beautiful the country looked. He said in reply — 'Finlayson, I would not give the most barren hills in the Highlands of Scotland for all I see around me.' ... In the morning we had a nice salmon for breakfast. The Captain seemed somewhat surprised and asked where the salmon was had. Oh, we have plenty of salmon was the reply. Have you got flies and rods, said the Captain. We have lines and bait was the answer and sometimes the Indians take them with the net, etc. No fly, no fly, responded our guest. So after breakfast we went to fish with the line, from a dingey. When we came back we had four fine salmon, but he thought it an awful manner in which to catch salmon."

And that is all about the salmon and the fly. At home on his native Scottish moors the captain was probably a fine fellow and a generous host. Out in the colonies, in command of a man-of-war, he was a sticky and disgruntled old character, determined to like nothing and enjoy nothing. It is the captain's ghost I summon now whenever I learn something new of fly fishing for Pacific salmon.

I am never quite sure about the captain. He probably has a tricky point of view. We might have got him interested in trolling a bucktail or polar-bear fly in salt water, with a five- or six-ounce rod and lots of backing; when a fresh coho ran off a hundred yards along the top of the water, then jumped like a bright steel spring, he would have had to admit the sport had something. I think he would have preferred the estuary fishing and perhaps compared it, not unfavourably, to Scottish sea trout fishing. But he would have kept a faint sneer in reserve for anything less than a fresh, bright fish from an upriver pool because that, after all, was undoubtedly the way he remembered it at home.

So the fish I really like to needle the captain about are those bright, early-running fish that one finds in the course of steelhead fishing, before the main run of cohos has left salt water. I hoped the captain was watching one day early in September three years ago when two really pretty cohos came to a twist of tinsel and a wisp of wood-duck feather on a No. 10 hook. He would have had to admit there was something pretty good about that.

Usually I forget about the captain when I'm catching steelhead. Those would be hard fish to explain to him — sea-running rainbow trout that are more like Atlantic salmon than trout. I fancy he would have argued the point, but there's not much doubt that he would have felt better about the whole Pacific Coast if he had known of them. Unfortunately he didn't, so I just leave him out of the picture, especially now that I fish a floating fly for them a good deal of the time; even the captain wouldn't expect Pacific salmon, or Atlantic salmon for that matter, to take a floating fly.

I suppose I drift a fly over some hundreds or even thousands of Pacific salmon every year. They're in there with the steelheads and the harvest cutthroats and every drift goes over them, whether one wants it to or not. At some time or another every one of the five species has taken a sunk fly for me, but until 1953, I had never had anything that looked like an offer to a dry fly, and I had never heard of any other fisherman who had.

What happened in 1953 wasn't so very convincing either. It would have been around the end of August and the big spring salmon were already in the Campbell River. I was drifting a hair-wing fly along the edge of the deep, fast run in the Lower Islands Pool. I liked the way the fly rode the current waves and couldn't help hoping it might stir up a fish from somewhere, so I kept plugging it up and letting it ride back. At around the tenth or perhaps the twentieth drift a great handsome, green-backed, bronze-sided fish rolled up. He was well over twenty pounds and his nose seemed to meet the fly so perfectly that I almost tightened; then I saw it, still floating, past his gill-covers, along the faint bronze of his side, past the anal fin, past the tail. I could even see it checked slightly and twisted in its float by the brush of his scales. I pitched it up there again a dozen times, maybe fifty times, but the big fish never rolled or showed again. In the end I went on. After all, Pacific salmon don't take dry flies; the coincidence of a roll like that with the float of a fly would be far less strange than an honest rise.

Just over a year later, on September 19, 1954, I fished a big, floating fly very carefully all up the long bar of the Main Islands Pool. It was a grey day, slowly clearing. The humpback salmon were already spawning and the tyee salmon were rolling regularly along the far side of the pool, very big and handsome with their red, gold, and bronze spawning-colours catching the light. A run of steelhead had passed through about a week earlier and I was hoping for stragglers or perhaps a big cutthroat or two. Anything can happen in the Campbell in September, but this time it didn't. When I came to the main run of current over the bar I had raised only one fish, a small one of fourteen inches or less; and the main run itself fished blank though I spotted the fly all over it.

When things are as quiet as that I usually search back down the pool with a wet fly. But I always ease up to the head of the main run before changing over, wedge myself in against the current, and start dragging my dry fly. The main run drops off quite steeply into deep water. I throw the fly across on a slack line, let it drift along the far edge of the run, drag it across as the line comes tight, then bounce it up against the current waves until it is over the shallow again, where I slack off and let it drift back in case anything has followed. Often enough this has produced a fish for me, anything from a two-pound cutthroat to a fifteen-pound steelhead. It produced this day. At about the third cast a big, bright fish flashed up on the shallow, grabbed the fly before it had drifted back two or three feet, and turned down with it.

I tightened on him and got set to let everything go. Fishing steelhead at that time of year I use 2x gut on a sensitive nine-foot, five-ounce rod, with a hundred yards of backing behind the fly line. There is no arguing with a summer steelhead's first run and if everything doesn't run free, 2x gut simply isn't good enough. But this time the explosive first run didn't happen. The fish was heavy and in control, but he only ran out ten or fifteen yards of line, then seemed to want to sulk.

I decided to stir him up, 2x gut or no, and he came rather easily, as though puzzled about the whole affair. I didn't want him too close to the rod top at that stage of things, so I eased up again. For a moment everything was quiet. Then he jumped, straight out, right up to eye-level, and not ten feet from me. As he hit the water again he took off, downstream and fast.

While the fish was in the air my eyes told me: coho, not steelhead. But I promised myself not to believe that, not even to think about it again until the fish was on the beach — if I ever got him there. But the run made me think. It went on and on, now very fast, now slowing a little, now very fast again, but always in a straight line, across the pool and down towards the big rock at the tail. And I could feel him shaking his head and twisting his body against the restraint.

With a steelhead I never move down the pool on the first run; if anything I move up, to keep as much water as possible between the fish and the rapid below. The steelhead usually change direction quite sharply, without slackening speed, somewhere in that first run, and jump two or three times. They come back after taking forty or fifty yards of backing, then run again, almost as far and fast. This run went on and on, far past fifty yards of backing. When the fish broke water below the big rock I judged I had less than twenty yards left on the reel and I started to edge down to follow. Then I realized he had turned.

He was a long time coming back — it was a long, long time before I could even see the backing splice again — but he came quietly all the way, right up to the shallows. There he jumped again, well clear of the water, but a tiring jump, and I found

I could roll him and make him flash with a little side pressure.

I had worked a few yards down along the bar as he came back, but there was still a good fifty yards to go before I could hope to beach him safely. I began to lead him down, trying not to look too closely at him in case he broke loose at the last moment. But I saw his tail several times in spite of myself; it was an unspotted tail, not a steelhead's tail at all. Near the end he made one more feeble run, but I turned it easily; he let me strand his head on a flat rock and I killed him.

There was no doubt then. The fish was a fresh-run female coho of ten pounds. For the record, she had thirteen anal fin rays, which means Pacific salmon, about sixty-five pyloric caeca, which means coho or sockeye, and twenty-two rakers on the first gill-arch, which means coho and not sockeye. The stomach was empty, there were no sea lice, and the colour was bright silver except along the back which was steel-grey.

I thought of the captain and wondered what he would have to say about it all. Then I remembered that up in his Highland fastness he had never heard of a floating fly. He was Roderick Finlayson's guest in 1845, only four years after George Pulman, down in Devonshire, had written the first record of the dry fly. It would have been a tough idea to put over to the captain, and I doubt if he could ever have grasped the full significance of it. But perhaps his ghost is piscatorially more advanced, and I hope it will be with me when I go up the river this year to look for another Pacific salmon with a floating fly. If they'll do it once, they can be made to do it again.

6
Along the Steelhead Rivers
(1970)

It is easy to explain the steelhead by saying that he is a sea-running rainbow trout. But he is also an extremely complex and adaptable creature, with many different patterns to his sea and freshwater life. At times he may even remain in the river and live out his life as a resident rainbow, but when that happens something has probably affected the influences of heredity, because there is no real doubt that the sea-running habit is hereditary.

The steelhead is also one of the world's truly great game fish, handsome and graceful in all his phases, strong, fast and violently active. He will take fly, spinner, lure, or bait readily and firmly. He is also big. Twenty-pounders are not uncommon on some rivers; the record rod-caught steelhead weighed thirty-six pounds, and John N. Cobb, presumably referring to net-caught fish in the Columbia River, says that extreme sizes reach forty-five pounds. Cobb also suggests that the steelhead got its name "from the hardness of its skull, several blows of the club being required to kill it when taken into the boat." This is in contrast to the several Pacific salmons, compared to which the steelhead has a much heavier and stronger bone structure.

The steelhead is a true trout, closely related to the brown

trout of Europe and to the Atlantic salmon. In habits he is closer to the Atlantic salmon and the European sea trout than any other fish, and the structural and other physiological differences between these species are very small; there is little reason to doubt that the steelhead and the Atlantic salmon shared a common ancestor in fairly recent geological times — probably about two million years ago, when the Arctic passages between the Atlantic and the Pacific were relatively open. It has often been said that the steelhead and not the oncorhynchids should have been called the Pacific salmon, and this makes good biological sense, even though there is evidence that the oncorhynchids themselves branched from steelheads at some evolutionary stage.

The native range of the steelhead is roughly the perimeter of the North Pacific, from southern California to Bristol Bay in Alaska and south along the Asian shore to Kamchatka Peninsula and the Okhotsk Sea. At one time the main centre of abundance was probably the Columbia River, and the mountain streams of the heavily forested Pacific slope seem to have provided ideal freshwater habitat. In its ocean life it mixes with the Pacific salmon stocks and ranges well out into mid-Pacific.

Because it differed from the far more abundant and obvious Pacific salmon runs, the steelhead was something of a mystery fish to the early settlers and fishermen and there were many strange theories about its origin and habits. All anglers agreed, though, that it was a strong and formidable fish, given to long, fast runs and mighty leaps that tested tackle to the utmost. Innocent trout fishermen, with the telescopic steel rods, nickel-plated multiplying reels, Colorado spinners, and worm hooks that were popular in the twenties, gave awestruck accounts of savage strikes, broken lines, broken rods, and giant, silvery fish leaping against the dark background of the rain forest. These fishermen knew they were out of their league and, as I recall it, seldom showed any inclination to go back for more. But there were also steelhead enthusiasts and experts who knew the runs and the rivers and the holding pools. All of those I talked to when I first came to the Pacific Coast, and some of them were very knowledgeable fishermen, were firmly of the opinion that the only practical way to take steelhead was on bait, preferably salmon roe. Spoons or spinners might take occasional fish, but real steelheaders didn't bother with them. Some of my experts had heard of steelhead taken on flies, but considered such accounts no better than wild tales.

If I overstate these views — and possibly I do because memory is a frail thing after more than forty years — it is not by very much, and certainly the impression I gained was that anyone who went after steelhead with anything except salmon roe could only be a wild-eyed optimist and a greenhorn whose ways would change. This unfortunate attitude was probably developed on the lower reaches of streams that often carried heavy loads of silt — those generally most likely to be accessible to early fishermen — and persisted as a matter of habit. It still persists among large numbers of fishermen; it has been very damaging to the quality and standards of the sport, unnecessarily destructive to steelhead stocks, and it has effectively downgraded the steelhead's reputation as a game fish. Steelhead can be taken quite readily enough under almost any conditions, except heavy flood, by spinners and lures of all descriptions. And they can be taken by orthodox fly fishing methods at all times of year, though winter fish are difficult to move to the fly under some conditions.

Over most of their range, steelhead divide recognizably into summer- and winter-running fish. Fish entering their rivers between May 1 and September 30 are usually rated as summer-runs, while winter fish run from approximately November 1 through April. The distinction is an important one, because true summer runs come in as green fish, with little development of milt or eggs. They mature in the rivers and usually spawn early the following spring. Winter fish have well-developed ovaries and milt sacs and, in spite of their late entry, are also spring-spawners. The two races are physiologically separable even in the premigrant stage. Summer-run parr and smolts have significantly larger quantities of fat in the abdominal cavity, among other differences, and this probably reflects a body design better adapted to withstand the long freshwater wait for maturity.

A few rivers have major runs that vary substantially from this pattern. The Cheakamus and Atnarko Rivers in British Columbia, for instance, have late winter runs that make good fishing in April and May, though the fish spawn very soon after coming in. It has been suggested that the famous big-fish runs that show up in the Kispiox, Sustut, Babine, and Morice Rivers in late September and October are really early winter fish. These streams are all tributary to the Skeena system, at considerable distances from salt water. Numbers of big steelheads are passing through the nets at the mouth of the Skeena in August, to my knowledge, and probably earlier; so it seems more likely that these late arrivals are summer fish that have taken their time in working up the main river.

Typical steelhead life histories are much like those of Atlantic salmon. The young fish spend from one to three years in the stream, growing to a size of four to ten inches before silvering up and going to sea in spring or early summer. They return to spawn for the first time after one, two, or three years of ocean life. A fair number of fish spawn a second time after another year of ocean life. A few spawn three times and very few four times. Seven or eight years seems to be about the maximum age, though it may be longer in the far north.

Size is certainly an important factor in determining migration times, whether from fresh water to salt, or from salt to fresh. Size is associated with growth rates and growth rates with water temperatures. In southern California streams, the numbers of yearling and two-year-old migrants from fresh water may be approximately equal, and grilse runs — that is, fish returning to spawn after only one year in salt water — are more numerous though two years of saltwater life is normal.

Working northward through northern California, Oregon, and Washington, two-year-old smolts increasingly predominate. In southern British Columbia, yearling smolts are rare, two-year-olds make up 40 per cent, and three-year-olds 55 per cent, a few fish delay migration to their fourth year. About 60 per cent of the fish return to their first spawning after two years in salt water, 37 per cent after three years, and 2.6 per cent after four years. The rather surprising effect of this is that the average size of the mature fish increases in the more northerly streams; in other words, the extra year of growth more than makes up for the slow growth of the early years. The average length of adult winter fish varies from about twenty-three inches in California to twenty-six inches in Oregon, and over twenty-eight inches in southern British Columbia. Dean River fish (in central British Columbia) and Skeena River (Kispiox-Babine) fish in northern B.C. clearly average a good deal more than twenty-eight inches, so it seems likely that the age progression continues as well.

Summer steelhead runs are among the most valuable game fish stocks in North America and probably the most vulnerable. The shrinking glaciers and the great rain forests fostered them. Each river's runs, and the races within the runs, developed slowly over several thousand years to fit the conditions they found. The conditions have been changed and are being changed at a rate far beyond the possibility of evolutionary adaptation. Dams and pollutants of many kinds have done their part, in some cases wiping out stocks forever. But there can be no real doubt that logging is the most serious factor of all, because it is so universal and so very badly done. Logging, as it is practised on the Pacific slope, produces winter floods and summer droughts; it causes silting of gravel beds, destroys shade, and raises stream temperatures. All of this is bad enough on the lower reaches of the streams, but in the headwater tributaries, where the loggers are increasingly reaching, it is immensely destructive. It need not be. Logging has to go on, but it can be controlled so that the small watersheds are logged at intervals in staggered settings.

The final threat to the summer runs may well be in the fish that the hatcheries are now able to turn out and bring back so successfully. To explain this seeming paradox in a few oversimplified words: hatchery stocks are specially selected for hatchery purposes — timing, rapid growth, early maturity, high fertility, resistance to hatchery fostered diseases — nearly all of which are totally unrelated to the factors of natural survival, environment, and competition, where the natural variability of the stock plays an all-important part. Variability within a species underlies natural selection — the survival, and procreation, of the fittest. Interbreeding might prove to be detrimental since hatchery fish have been artificially rather than naturally selected, bred for characteristics other than those which will make them fittest to survive in the wild. Restrictive timing, for instance, would reduce adaptability to natural disaster and perhaps favour early- or late-runners. There may be loss of resistance to natural diseases or parasites. The chances of artificial selection might even favour physiological changes, which could in turn reduce the ability to hold over to favourable spawning times. In any event, some modifications of a stock of long-proven viability must be expected. In 1969, some 65 per cent of the North Umpqua summer run was made up of hatchery fish. The question that suggests itself is how much damage will be done to the remaining natural stock by interbreeding, not only this year but in succeeding years, as the hatchery stock becomes progressively more specialized? The answers that suggest themselves are not good.

In the meanwhile, the North Umpqua remains one of the best and most beautiful of summer steelhead streams, and it has the tremendous asset of several miles of water restricted to "fly only." The strong flow of bright water is broken up into rapids and chutes by ledge rock outcrops, the pools are deep and long and hold fish well, and the fish themselves are usually responsive and in excellent shape. This is the territory of the Steamboaters, a dedicated group of steelhead enthusiasts who have their headquarters at Frank Moore's Steamboat Lodge. They mostly fish a floating line with about ten feet of intermediate tip that goes just under the surface when it is wet. The fly is kept slow by occasional mending. The most popular and most successful pattern is the Skunk Fly — red tail, black chenille body ribbed with silver, white hair-wing, and a black hackle — on a No. 4 or No. 6 hook. There are other local favourites, such as the Umpqua Special, the Silver Steamboater, and the Copper Colonel, but in my limited experience the Skunk outfishes anything you care to try against it, and my Steamboater friends usually admit to the same experience. The Steamboaters beach a lot of fish in a season, but rarely kill one, and in their short existence they have established a reputation for constructive conservation work.

A good number of northern California, Oregon, and Washington streams have good runs of summer fish. The Eel, Klamath, and the Mad in California; the Rogue, Umpqua, Deschutes, and others in Oregon; the Wind, the Kalama, the Grande Ronde (also in Oregon) and the Stillaguamish in Washington are all famous streams, several of them with stretches of "fly-only" water. British Columbia had the Capilano, the Seymour, and the Coquihalla, among others on the Lower Mainland, and the Stamp on Vancouver Island, but none of these can now be considered good or even fair for summer-run fish. The Stamp and the Coquihalla have the best possibilities of recovery. The Dean River, near Bella Coola, has been very popular over the past ten years. It is a big-fish river and quite a number of twenty-pounders have been taken on the fly there. The Skeena tributaries — Kispiox, Babine, and the Morice — have been well-known since the Second World War. The Kispiox particularly has attracted steelhead enthusiasts from far and near because of the size of its fish; the present rod-caught record of thirty-six pounds was taken here and twenty-pounders are, if not common, to be expected. Like summer runs elsewhere, the Kispiox fish come well to the fly — the record fly-caught fish is a thirty-three-pounder — but there is a great deal of hardware fishing. Fortunately, the use of salmon roe has now been forbidden on the Kispiox, the Morice, and their tributaries.

British Columbia probably has good summer-run streams still to be discovered, but the search is by no means easy. Clear tributaries of some of the big glacial streams of the mainland inlets are good prospects; some of the Nass River tributaries should have runs of fish as big as the Kispiox monsters; but these are remote places, hard to get to and hard to move around in when you do get there, so discoveries will not come easily or quickly. The west coast of Vancouver Island, which has many promising possibilities, has been disappointing so far. Access is difficult everywhere and good runs have a disconcerting and largely unpredictable habit of disappearing into lakes or canyon waters after a brief time in the more favourable pools.

Summer steelhead runs will never be found easily in country that is not fully opened up, because they so frequently pass through the lower reaches of a river to some remote tributary or to canyon water or to lakes. Their distribution still puzzles biologists and it has been suggested that falls or stretches of badly broken water that are passable only at low summer flows may have something to do with it. Probably there are other factors as well. Many streams that have excellent winter runs apparently have no summer runs at all. But in my experience a few fish do enter most such streams in the summer months. If one can learn their timing and discover their holding water, they can be found there season after season. If river or watershed conditions change to favour them, these "trace" runs might quickly develop to full scale.

Summer steelhead, though still far from maturity, are essentially non-feeding fish and they can become dour and difficult after they have been some while in fresh water, exactly as Atlantic salmon do. But summer temperatures make them active fish and there is nearly always some way to move them to a fly. I enjoy them most under low water conditions when they come best to a floating fly or to a slender wet fly fished just under the surface film. But they will come also to small wet flies fished well up in the water, to larger flies worked well down near the bottom on sinking lines, and even to flies fished run-of-the-river fashion in midwater. They are available to the angler over a long season under ideal conditions. If we had any sense at all, we would restrict fishing for them to fly only, but we might well carry that two steps further: limit fly fishermen to barbless hooks and encourage them to release all the natural stock fish they catch. I have yet to see a steelhead spawning area that was overpopulated.

Winter-run fish are a much more difficult matter than summer fish, especially in northern rivers where temperatures are consistently below 45° Fahrenheit and air temperatures are often much colder than the water. I have taken winter fish on the fly in water at 32° (actually below that on my thermometer), with anchor ice on the stream bed and the line freezing in the rings in the course of a single cast. But there is no doubt that the fish are slow and reluctant at these low temperatures and perhaps impossible to move to a fly unless they are holding in very favourable water.

Even under the worst of such conditions, though, winter steelhead can be readily taken by spinning or other artificial lures, provided they are properly fished. There is only one way to fish them properly, and that is as slowly and deeply as possible without hooking bottom. The fisherman must know his river, both top and bottom, and where the fish lie. In this, I believe, the spinner fisherman needs even more skill than the fly fisherman, but he has the important advantage of being able to cast farther, it is much easier for him to get his lure down quickly and keep it near bottom, and the flash and flutter or vibration of the lure itself are probably important advantages.

The bait fisherman has all these advantages, plus the advantage of offering something that looks good, smells good, and actually is good to eat. He has a much better chance of tripping the fish's semi-dormant feeding reaction; the degree of deceit is minimal. Bait apologists are quick to point out that a poor fisherman doesn't do much better with bait than with anything else, and this is true. Even bait won't catch fish unless you put it where the fish are. But a good bait fisherman — and some are very good, indeed — is altogether more deadly than any sportsman needs to be. A friend of mine, fishing salmon roe, hooked and released the same fish three times in twenty minutes.

I have had much pleasure in fishing spinners for winter steelhead and the only reason I no longer do so is that my fishing time is limited and I prefer to spend it with the fly. One fishes exactly as for Atlantic salmon on European rivers, slow and deep, covering the water thoroughly, and holding or hanging the lure where possible. Winter steelhead will respond readily to the same Phantoms and Devon Minnows that please Atlantic salmon, but there are a number of specialized lures that are also very effective, among them the Tee Spoon, the Cherry Bobber, and the Spin-'n'-Glow. I very much prefer rotating reels, such as the Hardy Silex or the Swedish Ambassador, to the fixed drum reels, because I believe the former are more sensitive in fishing the lure and infinitely more satisfactory once the fish is hooked.

Winter fly fishing is largely a matter of getting the fly down deep, and the modern fast-sinking lines are a great help in this. Years ago I used double-handed rods, sometimes as long as fifteen feet, which were a help in controlling the silk lines of that time. But I now prefer a single-handed rod that will throw a No. 9 or No. 10 line; these lines are heavy enough to handle flies up to 2/0 or larger in comfort and can be thrown by a glass rod of nine feet and five ounces or less. For larger flies, which may serve a purpose under very cold conditions on big waters, I should still use a double-handed rod. Perhaps it is a matter of confidence; one feels insignificant enough on a mighty stream like the Thompson in midwinter, even with a fifteen-foot rod.

Winter fly fishing has developed a lot, in both techniques and popularity, over the past ten or fifteen years. Some members of the fly fishing clubs in Oregon, Washington, and British Columbia are consistently successful on winter streams, and every season a few more take up the challenge. We all realize, I think, that the most important factor is in finding winter holding-water that is favourable to the fly. After that it is a matter of getting the fly down there, close to the fish, and giving him time to react.

With water temperatures above 45° Fahrenheit, as they are likely to be in April and May, and earlier than that on some streams, there is very little problem. But I am not at all convinced that the deeply sunk fly is essential, even in really cold water, so long as the air is warmer than the water. Atlantic salmon can be moved to a fairly small fly, close to the surface, in low water temperatures — on days when the air is warm. We have all had winter fish come up to flies that were fishing much less deeply than we intended. Last winter I used a floating line with a No. 6 fly on the few warm days I was able to get out. The results were inconclusive; I hooked two or three fish, all of them in very shallow water. But I believe the theory deserves much more extensive testing.

We have been far too slow to appreciate the outstanding qualities of the steelhead as a game fish, much too willing to meet the difficulties he presents with crude and unsatisfactory solutions. There are signs of change in this. We have been slow, also, in investigating the variety and complexity of life history patterns that make the steelhead runs, and unimaginative in our limited attempts at management. The steelhead stocks of the North Pacific make up the world's only major population of a true sea-ranging trout. This is a precious inheritance. It needs cool, clean waters and good gravel beds to come home to; it needs stable flows and productive stream beds to nurse it through the freshwater years until it is ready to go to sea; it may need protection in the ocean years as well — if not now, at some later date. All these things can be provided in good quantity by wise planning.

The angler, for his part, should respect the fish by accepting the challenges he offers instead of insisting on easy solutions. And he should be satisfied to limit his kill closely until natural stocks repopulate every square yard of suitable spawning area, whether man-made or natural. Those great bars of muscle, steel-grey and silver, square-tailed and heavy-shouldered, have more important duties than filling freezers or posing for posthumous pictures.

7
Steelhead Angling Comes of Age
(1975)

The greatest single change in steelhead fishing over the past ten or fifteen years has been in the enormously increased acceptance of fly fishing as a practical method, not only for summer-run fish but almost equally for winter fish. This shift of attitude is all the more surprising since it has come about in the face of an ever increasing intensity of fishing pressure on streams everywhere.

There have always been a few winter steelheaders who fished the fly with confidence and success, but they were few indeed until after the Second World War and even in the 1940s and 1950s many good fly fishermen felt that it was only logical to turn to other methods during the winter months. Every winter brought a few converts — those who successfully beached their first fly-caught winter steelheads and so gained the necessary confidence to go on. This gradual recruitment has had its effect in new thoughts, new flies, new gear, new methods — and still more recruits.

There can be little doubt that the development of plastic sinking lines has been the greatest single factor in encouraging the winter steelhead fly fisherman. Before that there had been lead-core lines and shooting heads, but I never met a fisherman who enjoyed using them. Most of us used old silk lines, cherished because most of the dressing was worn off them, so they soaked up water and sank quite readily. I used to treat mine with linseed oil to reduce freezing, but apart from that the only concern was to keep them away from grease in any form. Even when well-soaked, these lines did not sink very readily. It was necessary to help them down by upstream casting, by mending and steering them around in the currents; and their large diameter tended to lift them rather quickly when the full current caught them.

Even so, they caught fish in the coldest weather and water and I am inclined to think that their relative liveliness may have had some advantages. I used long rods with the silk lines, seldom less than ten feet and often up to fifteen feet and these were a great help in guiding the fly down and holding it where and how it should be fishing.

The modern high-density fly line sinks fast and convincingly; the fisherman is readily satisfied that he is getting his fly down where the fish are if only because some brief inattention is likely to wrap it around the rocks on the bottom and prove the point. The line handles and casts comfortably enough; in a heavy wind its weight and small diameter can be a significant advantage. Its fast sinking quality makes it difficult to mend and almost impossible to control while the cast is fishing out, but the idea that the fly is sweeping around close to bottom is compensation enough for these disadvantages.

Or is it? A good fly fisherman wants control of his fly as it fishes and soon starts to look for ways to achieve control. Even when we were still using silk lines to get down some of us spliced a length of undressed silk ahead of a length of floating line, in approximation of today's "wet-heads" or "sinking tips." We could then mend line throughout the cast and there was excellent control, especially with the long rods. There was also the realization that winter fish often take quite well up in the water and that they will sometimes follow a fly around, making little plucks at it that can be faintly felt

— a situation where control may make all the difference between a fish securely hooked and nothing at all.

Considerations such as these have long since turned the sophisticated winter fly fisherman from the simple techniques of high-density lines and weighted flies to more interesting things — even to floating lines and small flies.

I have not personally fished enough in the past ten years to prove anything, even to my own satisfaction. I have hooked a few winter fish, always in shallow water, with a floating line and a small dark fly. I have occasionally dragged a largish floating fly, without result, over places where winter steelhead might have been lying; and I have wished for time and opportunity and ambition to do more than this. Other men have done better and have developed techniques of their own in which the high-density line no longer plays much part. One friend tells me he has taken winter fish with a nymphing technique, which is reasonable enough if one has the fortitude and patience. I have more than once found stonefly nymphs in the stomachs of winter fish.

In spite of all the thought and ingenuity and skill expended in the last ten or twenty years, I do not think there has been any major advance in wet fly patterns for steelhead. The winter fisherman still has a general preference for red and orange patterns and there is still the contrast between these bright beauties — Skykomish Sunrise, Lady Godiva, Comet, Umpqua Special, Prawn Fly and many others — and the dark patterns that catch fish equally well: Skunk Fly, Golden Demon, Boss Fly, Purple Peril. Like most other fly fishermen, I fuss over patterns and take great pleasure in them, but do not believe they make much difference to the steelhead. Weighted flies may sometimes be an advantage in that they get down quickly through the water, but I think the freer movement and more natural sinking rate of the unweighted fly are preferable most of the time. I sometimes remind myself that bait and spin-casters catch more fish every season with fluorescent wool wrapped on a hook than we fly fishermen do with our most elegant conceptions and our deepest probings of steelhead psychology. I find the exercise a healthful one since it reinforces humility and bolsters confidence at one and the same time.

Still on the subject of gear and equipment, it seems worth mentioning today's fiberglass rods. Most westerners — in fact most North American fly fishers — prefer single-handed rods. A nine-foot five-ounce fiberglass rod can be made to throw a No. 10 or 11 line and a big fly. In the past I have used cane rods eleven feet long and weighing twice as much to get the same result, and rods of ten feet or more weighing eight ounces were common. There can be little doubt that the lighter fiberglass rods have played a part in increasing the popularity of winter fly fishing. There is now a Fenwick graphite rod ten and a half feet long, throwing a No. 9 line, that weighs only four and three-eighths ounces; I can well imagine that this might make a very fine winter steelhead fly rod, if one had the money to risk on it.

In summer fishing I think the major advance of the past few years has been the increased acceptance of the floating fly. It may not be productive on all streams, but on most it will produce as much if not more action than a wet fly, and I know of dry fly enthusiasts from California to the Kispiox, from Washington's Wind River to the Dean and the Morice and the Copper in central British Columbia. Summer steelhead fly fishers are still immensely conservative; far too many still put their faith in ordinary wet fly fishing when they would be further ahead if they worked on variations of the greased line technique. I feel confident that a floating line, fished with skill, will always do better with summer fish, and it is by far the more attractive method. Under really low water conditions it may be the only effective method, apart from the floating fly.

So much for gear, equipment, and method. The last ten years or so have established the point that hatcheries can produce steelhead trout. Whether this is a good thing or a bad thing remains highly debatable. What is clear is that there has been an orgy of hatchery production, especially in the State of Washington, at a time when nobody had a very good idea of what the results might be. Certain dangers and drawbacks are now apparent, and steelheaders in California, Oregon, Washington, and British Columbia are beginning to ask serious and searching questions.

There are places where hatcheries are pretty well essential if there is to be any fishing at all; where a stream is totally blocked by a major dam, for instance, a hatchery provides the cheapest form of "mitigation" — a weasel word much loved by engineers and developers that means "an inferior substitute for what has been destroyed." A hatchery by itself will do nothing for steelhead, since they need at least one year of stream-rearing capacity, so this calls for rearing ponds and artificial feeding. When the smolts are finally released to go to sea, they are expensive little fish. Since only one in ten or so will return as an adult steelhead, any hatchery steelhead that ends up in an angler's catch is likely to have a cost of at least eight or ten dollars. This is considered economically sound and most hatcheries are able to show a favourable cost-benefit ratio.

Among the more obvious disadvantages of hatchery-reared stocks of steelhead are the following: they fail to distribute through the stream system, generally passing straight through to the hatchery; they tend to return within the same short space of time, thus reducing the length of the season; some stocks mature early, at smaller size; fishermen often claim that the hatchery fish perform poorly when hooked. Less obvious and more serious disadvantages are the following: hatchery stocks are genetically inferior and much more susceptible to disease; this inferiority may be continuously progressive; hatchery stocks cannot withstand pollution stresses as well as wild stocks; since the individual fish are larger when released than wild fish of the same age, the hatchery fish provide excessive competition, to the detriment of wild stocks;

some hatchery fish may spawn naturally with wild fish and impair the genetic quality of the wild stocks.

Hatchery stocks may also throw out the very complex balances and interrelationships of a stream. There is a tendency in some stocks for significant numbers to remain in the stream, providing excessive competition for the limited supply of natural food. Winter steelhead plantings approaching the limit of potential (the so-called density barrier) may interfere with summer runs and vice versa. Interspecific and even intraspecific predation may also significantly disturb stream balances.

Hatcheries have unquestionably produced more fish, even when one allows for the wild stocks that may have been depleted or wiped out. They have also attracted many, many more fishermen, both to the inferior type of sport they themselves provide and to the rivers in general. This may or may not be a good thing and it may or may not last, but it is certainly one of the most significant developments of the last ten to twenty years. Naturally enough, all this taken together has produced a very important reaction. Steelheaders, in California, Oregon, Washington, and British Columbia have become considerably disenchanted with the hatchery idea and highly suspicious of its long-term effects.

This has led to a deep concern for wild stocks and for rivers capable of maintaining wild stocks. I cannot imagine a more valid or more sportsmanlike concern. Most fishermen are agreed today that there is room for hatcheries and hatchery stocks — on the destroyed streams, the streams incapable of producing and maintaining a wild stock. Equally, they are saying: keep the hatcheries and hatchery plantings away from the wild streams, those still capable of maintaining a natural stock in good health and reasonable abundance. More directly, many fishermen conscientiously and carefully release every natural stock fish they catch. Perhaps this is a small thing, but symbolically it is very large. It expresses the will to save the fish, to respect and cherish a magnificent animal in its natural state.

In Washington State it may be that natural stocks are almost extinct — yet there must still be many streams where cessation of planting would permit them to recover, over a period of years. In Oregon and California natural stocks may be endangered, but there are still many streams in condition to ensure their future.

In British Columbia natural stocks are not yet endangered, but the threat of hatcheries hangs heavy over us. It may well be that our wild streams and natural stocks, because of their low productivity, will prove especially vulnerable to heavy plantings of smolts. But it has not happened yet and one can only hope that the authorities will use more caution and restraint than has been used elsewhere.

It should be axiomatic that all hatcheries and all hatchery plantings must be managed in such a way that they do not deplete wild stocks. There is, I believe, a great deal of room for further research and experiment into methods of stream improvement. Where supplementary artificial culture is necessary there should be much greater efforts to reduce artificiality; abundant and economic fry production, for instance, could be achieved on many streams by spawning channels and I feel that all fry-to-smolt rearing should be in flow channels at relatively low density so that the young fish have a significant proportion of natural food. Estuary rehabilitation or improvement is a possibility at many stream mouths.

These, and others like them, are the sort of things I hope that steelheaders will press for in the next decade. I hope and believe that steelhead fly fishers will continue to grow in both numbers and skills. I hope that the catch-and-release philosophy and use of the barbless hook will continue to spread among steelheaders of all kinds. I hope, too, that we can all continue to grow in stream ethics, in conservation ethics, and in respect for the quiet and gentle traditions of our sport.

8
They Pass in the Night
(1955)

Hunters and fishermen were the first explorers, and the joy of discovery has never faded in either sport. The hidden lake where the trout run big, the forgotten lagoon where the wild geese pitch, the sunken rock in the bend where the salmon lie, the bowl in the mountains where the rams shelter and grow big, all these are remembered triumphs far beyond any mere counting of slain fish or birds or other creatures. It is not necessary to be the original discoverer, though that is perhaps the brightest pleasure of all; it is enough to feel the discovery, to have come to it by some exercise of wisdom and woodcraft, and to know that it is shared by only a few others.

Living in comparatively new country I have had my share of discoveries, from steelhead streams to brant beaches, from goose meadows and bear slides to lost lakes and secret pools. Many have been handed to me on the silver platter of local superstitions that declared: "The salmon won't take inside the river"; "Those trout never come up to a fly"; "You have to go deep with spinner or bait to catch winter steelhead"; or some other sad theory as unlikely and untested. Discovery goes on and on. There is always a new place for a trout to rise in the most familiar stream, a new piece of cover in the most familiar woodland from which a ruffed grouse may flush.

The most satisfying discovery of my life was in just such a familiar place as this — the river I have lived beside for over twenty years. It was gradual, so there is a story to it. It was important — nothing less than an unsuspected run of desirable fish. And it is recent — so recent that it may not yet be complete. But it has grown and been repeated through enough seasons to be reckoned a sure discovery and not just a freak happening.

Summer steelhead are disappointingly frivolous in their choice of river to run to. One stream may have an excellent run; another stream a few miles away, apparently just as suitable, will have no run at all, even though it has a fine run of winter fish. Most of the streams along the west coast of Vancouver Island have summer steelhead runs, but few of them except the Stamp and Ash Rivers, near Alberni, are readily accessible to anglers. Along the east coast of Vancouver Island, where most of the streams are accessible by road, summer steelhead are scarce enough to be reckoned almost non-existent. Along the mainland coast inside the Island, you will find splendid runs in some streams, none at all in others nearby.

When I came to live beside the Campbell I accepted in good faith, or fairly good faith, the strong local opinion that the river has no run of summer steelhead. The Campbell is, after all, a very short river in terms of migratory fish — less than three miles from salt water to the drop of Elk Falls which effectively bars all passage. Its winter run of steelhead is, or was, magnificent, and its summer runs of cutthroat were good enough to keep fishermen active and, seemingly, to ensure that runs of other fish would not pass undetected.

My first discovery in the Campbell was a run of small steelheads, mostly two- and three-pounders, in late April and early May. There were larger fish among them, up to six or seven pounds, but a four-pounder was a big one and all the fish of good size were almost ready to spawn even though they were freshly in from the sea. By mid-June most of them had spawned and disappeared from the river. It was not a true summer steelhead run in any sense, and I knew the fishermen who might have come across it before me would probably have called the fish river rainbows or even sea-run cutthroats, and let it go at that.

Once, in late November or early December, I caught two mended steelhead spawners of about eight pounds. This was surprising and I remember saying emphatically at the time that the fish must have run in during the summer months and spawned late in the fall, probably in October. But I wrote them off as chance strays that might have come in with a school of salmon, and still did not realize that I should start looking for a summer run.

All this was before the war, and I too often neglected the river for saltwater salmon in the summer and fall months. But in August and September I always used to go up to the big pool at the foot of the canyon to find the harvest cutthroats. They were fine fish, but hard to reach in the big, wide slick at the tail of the pool, and cautious takers in the bright sunlight. I fished for them with a well-greased line, a long leader tapered to 2x or finer, and slender, lightly dressed flies that fished just under the surface film.

The method worked well for me and often enough I came home with half a dozen lovely three-pounders and a feeling I had earned every one of them. Occasionally, especially down towards the very tail of the pool, I broke off a fly in an unusually solid strike. This aroused my curiosity, but I could usually account for the trouble by blaming the angle of the take or some fault of my own in responding to it; I was fishing close to the limit of the leader's strength in any case, but I was unwilling to increase its size because the big cutthroats were cautious enough about 2x in that smooth and brilliant water.

How long this might have gone on, I am not sure. Once or

twice I caught a glimpse of a fish that broke me and I know I suspected there might be cutthroats twice as big as any I had killed. Probably I suspected, too, that there might be a strong steelhead among them, but it never struck me that this might be more than a matter of obscure chance, not worth calculating in my methods or tackle.

Then, one early September day in 1939 or 1940, a really big fish took my fly well down in the tail of the pool. It was a long cast and there was a good belly of line floating on the water between me and the rise, so I left it there and let it draw the hook home. He ran the moment he felt it, straight upstream and into the deep water between the rock walls of the canyon. My reel had a light check, I had plenty of backing, and I made no effort at all to control him. In the canyon he jumped twice and I saw he was really big — he would have been big even for a winter fish.

It would be silly to say I controlled any part of his performance from there on, except the very end of it. Fortunately he was fast and violent, swimming and jumping himself out, crossing the pool several times but always holding well up in it. When he tired, I held his short runs gently, turning them back again and again to the shallow water on my side of the pool, until at last he lay long enough on his side to let me slip a finger in his gills. He weighed sixteen pounds and is still the biggest summer steelhead I have ever caught in the Campbell.

This gave me rather different ideas about the big fish that had been breaking my 2x gut on the strike. But I was still inclined to write him off as a lucky encounter, a freak fish, until I went up to the canyon a few days later. There, hovering over a shallow ledge of rock in bright sunlight, I saw another perfect steelhead of at least twelve pounds. The fly I put over him only served to scare him off the ledge, but I was just as sure of the identification as I would have been if I had killed him.

The following season I caught a twelve-pounder, also on

2x gut and a No. 6 fly, in a big pool farther down the river, and two nine-pounders in pools near the islands. Then I was in the army and missed the fishing for several seasons. But my friend Tom Brayshaw managed to spend one or two of his summer leaves by the river and told me that he managed to find a few steelhead each time — occasional fish, one in this pool, one in that, never more than one in a day, or several days on the river; but enough to show that some fish ran in every year.

It is often hard to say at just what point a discovery is "proved." But proof is a very important thing if the discovery is to be a real satisfaction. I remember trying to prove that tyee salmon would freely take a cast spoon or spinner in fresh water; the first fish I caught was hooked just outside the mouth; the second took just after I had set the rod down in the boat and rowed a few strokes; the third was in tidal water; then one did everything right, but it still had to be repeated two or three times and preferably over two or three seasons before it was securely proved. It's the same way with new steelhead lies. When one catches a fish in an unusual place, it may mean much or little; he may have been a traveller or an individualist, or some chance height of water may have held him there. Two fish on different days from exactly the same lie begin to mean something, three, and the lie is well proved; it will always be worth looking for a fish there unless something happens to change the river bottom or the current flow.

It seemed to me that I wanted something more than occasional scattered fish to prove my summer run, and I was reluctant to count the smaller fish of two to four pounds which had spent only a year in saltwater feeding, because these seemed more closely comparable to migratory cut-throats than true steelhead. I found myself hoping for two or three fish in one day, all of them over six pounds. That, I thought, would have real meaning.

Shortly after the war a dam was built on the Campbell and

the summer flow of the river was greatly increased by the flow from the turbines. This meant that migratory fish, especially cutthroats, began to hold over in some of the shallower, swifter pools they had formerly passed through, and I was able to search for them with a dry fly. I developed the habit of using a rather soft, easy nine-foot rod, a small reel and 2x gut with a large dry fly. The cutthroats came short to it sometimes, but they moved to it more readily than to anything smaller, and if I could not hook them I knew where to try for them with a smaller dry fly or a wet fly.

It crossed my mind that the increasing summer flow of the river — it has increased every year since the dam was put in and is still increasing — might build the numbers of summer steelhead and might also cause them to lie in the more open water below the canyon. But I was not wise enough to take any special precautions. One bright August day I waded into a favourite pool below a long, shallow bar with my little reel and my 2x gut, and began to look for the big harvest cutthroats. The hair fly bounced cheerfully down the first little run below the bar, a good fish rose to it and kissed. I threw it up there again, a silvery four-pounder grabbed it and went off like an explosion, so fast and fiercely that I was left fumbling, with only sense enough to leave the reel alone. The first run was upstream and across, towards deep water, but he turned from that suddenly, without coming back, and went full speed for the rapid below the tail of the pool. I tried to slow him a little and the leader broke at the fly.

I felt very foolish, but the warning was not enough. I put up another fly on the same leader and threw into the next run. Another four-pounder rose beautifully, head and tail, and I hooked him solidly. He fought like the first fish, but a shade more slowly and I managed to net him at last. I clutched net and fish against my waders and stumbled ashore to finish him off.

Then I took stock. This was a steelhead, bright and fresh

from the sea, as I was sure the first had been. These might be the only two or the pool might be full of them in all sizes up to ten or twelve pounds. All the leaders I had with me were tapered to 2x, and my little narrow-drum reel held less than fifty yards of backing. The only comfort I could find was in the fact that I was working up the pool, away from the fierce pull of the rapid at the tail. I decided there was a good chance I could handle whatever came up from there on and went out to try again.

The next run was heavier, more broken water. I drifted the fly down it half a dozen times without a response and tried one more drift before moving on. An enormous head, back, and tail rolled slowly up in the broken water and the fly disappeared. There was just time for me to realize that I was afraid to set the hook; but I suppose I did set it, because in the next moment the poor little reel was screaming its heart out until the fish jumped clear out and flopped over, forty or fifty yards upstream. There was no pause with the jump. He turned and came right back, fast, then ran straight across and jumped under the far bank, cutting the water upstream again from the jump until there was a great sweeping belly in the line. From there, still very fast, still without a pause, he came back within twenty feet of me. I caught up somehow, tightened again, and that was all he needed. He ran downstream this time, jumping wildly. The reel was screeching unpleasantly as the short turns of backing stripped from near the core and I tried to help by pulling off line with my hand. But the extra strain was too much and the leader broke again, without even a jerk. I could see the metal of the drum through the turns of line that were left, so it would have broken anyway I suppose. But I was shaken and angry. For the first time in my life I knew I was out of my league on my own stream.

A wiser man would have gone home to pick up heavier leaders and a better reel. I couldn't wait for that. I cut about four feet of taper off my nine-foot leader, increasing its breaking strain from two to about five pounds, then cut five feet off another leader and knotted the two together so that I still had nine feet of gut between the line and the fly.

I was at the second main run now, probably the best holding place in the pool at that height of water. I saw a big fish roll and set the fly just above him. It drifted over without response and I let it come, as I always do for steelhead. It must have been twenty feet below him when he came back fast, along the top of the water, and had it. I set the hook and gave him all the argument I thought the strengthened gut would stand. I suppose he was less wild than his cousin, though it didn't seem like it at the time. In the end I worked him back along the bar, beached him, and killed him, a perfect bright fish of seven pounds. And in the main run I missed two good rises, then hooked and killed another perfect nine-pounder.

The next day I was back there with all the right gear. The only fish I could find were two comparatively quiet cutthroats of three pounds, though once I thought a larger fish followed the fly back.

All that was two or three years ago. In the summer months I fish the same soft, easy five-ounce rod, with a free-running reel that has a fairly wide drum; and I usually trust to 9/5 or 0x leaders unless the day is very bright and the water very low. There had never been another day quite like that one, but I expect to find steelhead up there and every so often I do find them.

Perhaps it is too much to say that this represents a run of fish. When people ask me if there are summer steelhead in the Campbell, I still find myself saying: "Well, it's not exactly what you'd call a run. Just an occasional fish. Nothing you'd really go out and fish for." But I do go out and fish for them, or at least with a hope in my heart that my flies will find them. And every year I do find them, perhaps a little more certainly and a little more often each year.

And how did it happen that for nearly twenty years I let

them swim right past my house without noticing, the best fish of all, summer steelhead, neglected as though they were bullheads? Well, that's something else again. Perhaps, if there were more than the very occasional fish I found, they slipped through and up into the pools of the deep canyon below the falls without waiting to be found; I often wish I could have an infallible count of the fish that pass my house, upstream or down, each night. Perhaps the increased summer flow since the dam was built has encouraged them to be where I can find them more easily. But I would rather believe that the run is slowly and surely building up, that there were originally just a few strays and the new summer flow is giving them a chance to increase to a full-scale run.

If that's the way it is there will come, sooner or later, another day in the pool below the bar when I shall have the right reel and the right leaders and plenty of backing, and the fish will be fresh in from the sea. When that day comes I'll settle for three ten-pounders, and it will be a strong fish that breaks me.

9
Fascinating Challenge
(1971)

Pacific Coast winters are relatively mild — but only relatively. Snow and cold are common enough and a little way back from salt water the lakes freeze over. By January, temperatures in most of our British Columbia streams are running in the mid-thirties and sometimes lower. When it isn't cold it is usually wet, raw, and windy. The brush along the streams is thick and heavy and soaking wet, the river rocks are wet and slippery and the wind drives sudden showers of cold water out of the sodden evergreens. It is a well-dressed angler, closely zipped or buttoned, who can remain dry through such a day, even if he avoids the elementary error of falling in the river or filling his waders by some other incautious move.

These are sufficiently daunting conditions, I suppose. In the very coldest weather it would take only a few more days or a few more degrees to ice the streams over and stop fishing altogether; and nothing is colder or wetter than winter rain from the North Pacific, unless it is the winds and rain of Tierra del Fuego. But the fish themselves add still another dimension to the challenge.

They are splendid creatures, these winter steelheads fresh from the sea: thirty-inch bars of solid muscle, with steel-grey backs, and pale, clean bellies. On their sides, just above the lateral line, the blue-grey of the back breaks sharply to gleaming silver, sparsely spotted with black. Some stocks are thick-bodied and powerful, others slender and swift; all have strong square tails with thick "wrists" and heavy spotting. When they have been in the river a little while the deep ocean colours darken and change to the protective shadings of the river shallows, and some fish plainly show the rainbow's rosy stripe of maturity. They are the angler's dream trout, except for one thing; when they come into the rivers they are no longer feeding.

Fortunately this does not mean that they are beyond temptation. They do respond at times, both to the natural drift of natural food and the calculated drift of angler's bait or lure. But they are not actively feeding, they have no need and they will not go very far out of their way to take offers. In my experience, well over 90 per cent of winter-run steelheads have empty stomachs when caught, and even the exceptions show only casual sampling, perhaps a single insect, one or two salmon eggs, or a few preened-out merganser or goldeneye feathers. This reluctance is further compounded by the very low temperatures of the winter rivers. Between 50° and 60° Fahrenheit trout are very active; between 40° and 50° usually much less so; in water below 40° they are slow and lethargic, little inclined to move at all — in fact, they are very nearly hibernating.

This, then, is the sum of the challenge: rough and varied winter weather, the cold of the winter rivers and great, handsome silvery fish that have no practical use for anything a fisherman can offer them.

I came to winter steelhead fishing first in my late teens and the challenge to me was not in the winter weather, nor in the size and power of the fish, but in their reputed reluctance. Hard-nosed winter steelheaders were few in those days and the average fisherman I met considered the winter steelhead an awesome mystery, a fish too strong and tough for ordinary tackle and practically impossible to catch except on salmon roe. I was enough of a fisherman to doubt that and I wasn't about to use roe anyway, or any other kind of bait.

The answer was simple and one with which I was already familiar from a long apprenticeship on slow, quiet Atlantic salmon streams in southern England: it was essential to fish as slowly as possible, as deep as possible without hooking bottom, and to search the water thoroughly. Even today, some forty-five years later, I know no more than this about catching winter steelhead. There are better and easier ways of doing it now, and gear better suited to the purpose, but that is all.

One of the first winter steelhead I took was on a bright-pink artificial shrimp made of celluloid. The shrimp was a flimsy, hollow thing, too light to cast easily and completely un-lifelike in the water. I did hook a second fish on it before my confidence faded completely and I settled to that favourite Atlantic salmon lure of European waters, the Devon Minnow, of which I had a good supply. These are fast spinning shells, shaped like small fish with two propeller-like fins. They are available in many sizes and many different materials — aluminum, brass, vulcanite, wood, and now plastic. They cast well and spin very evenly and smoothly, even at slow speed in the gentlest current. For ten years I used little else for winter steelhead and took up to fifty fish a season without much difficulty; but there were few other fishermen in those days and I was fishing unspoiled streams.

Since the Second World War fixed spool reels and monofilament nylon lines, to say nothing of improved cold weather clothing, have made winter steelheading a popular winter sport. Every river within reach of a road has a full quota of enthusiastic anglers most days of the week, many of them out before dawn and waiting on the pools. But the conditions are the same and the fish are the same; it is still essential to fish slow and deep, still essential to know where the fish are and steer the bait or lure accurately to them; salmon roe remains the most popular and by far the most deadly bait nearly everywhere, but it is less than ever necessary. Modern lures like the balsa wood Cherry Bobber and, even better, the Spin-'n'-Glow with its rubbery fins and eccentric drift, are extremely effective and much easier to fish than the old-fashioned spinners and spoons. The problem is to get deep, right down to the fish without hooking bottom or snags — and where the quietly wintering steelheads lie, there are usually snags or rocks or both.

Hooking bottom often enough means lost gear and, more important, lost fishing time; rigging up again with cold hands can take time. So the good steelheader knows the bottom of his river almost as well as he knows the surface, and he is constantly dreaming up new ways to save gear. In British Columbia many fishermen use bobbers, adjusting them so that the lead weights bump along the rocks or gravel of the bottom while the lighter bait or lure sweeps along just above. With this rigging it is possible to cast upstream and get a long, reasonably safe drift down past the fisherman.

All fishermen talk endlessly about gear, none more so than the steelheader. Few of us, I think, really like the fixed spool spinning reel once a fish is hooked. Control is uncertain and uncomfortable with a heavy fish, because there is none of the direct touch of the conventional revolving drum reel. Many British Columbia experts use the famous English Silex reel and a long limber rod (ten feet or more) for bobber fishing, and it is hard to improve on this. For lure fishing I prefer a somewhat shorter and stiffer rod and a well-known Swedish casting reel that is superb for both distance and accuracy, though still not as satisfying as the Silex for playing a fish.

Individual preferences such as these would be quickly shot down in any group of experienced steelheaders, some of whom would certainly express and justify other preferences, then prove their soundness by going out and catching more fish than me. My own position is weakened, too, by the fact that for the past twenty-five years I have hardly ever fished with anything but a fly for winter steelhead. The fly is not as effective as lures or bait, especially when there is heavy competition, nor is it as easy to fish. But here, too, modern gear is a big help. Glass rods are more powerful for their weight than cane and make it easy to throw a big fly single-handed. Plastic lines of varying density enable one to get the fly quickly down to the fish and they pick up much less water than the old silk line, reducing the problems of freezing in the rod guides and the reel.

In spite of its difficulties and relative ineffectiveness under most conditions, more winter fishermen are turning to the fly every year. The advantages are many: there is the simplicity of the gear first of all, the pleasure of casting, the problems of wading, the satisfaction of the heavy strike straight to the hand, and the freedom of the fish to put on his best show. Most important of all, perhaps, is the fact that the fly rather seldom hooks bottom and even then will usually come free. A more practical advantage is that the fly fisherman can fish pockets and runs in the fast water between the pools where other fishermen seldom bother to go. Such water is not always productive, but it can have its pleasant surprises.

What are the great attractions of winter steelhead fishing? One doesn't go out there just to be miserable. The fish are there, of course, big and beautiful and not too easy to catch; some of them will run a hundred yards of line off the reel and jump three or four times as soon as you hook them. All this is excuse enough, but one goes for more important reasons. I go because I like to be out along the winter rivers when the alders are leafless, the sun is low, and the woods are quiet. I like the dark and heavy water, the worn remnants of the salmon runs, and the gulls and mergansers and golden-eyes that are using them up. I like the icicles on the brush at the water's edge, the bright bark of the red osier, the little grey dipper flitting about the rocks, in and out of the water, impervious to cold.

I am not miserable, nor even uncomfortable usually. I wade carefully and fish carefully, working my fly deep and slow on the cold days, trying to hold it and hang it over the likely spots. I watch and see things as I fish, ducks passing up and down the river, a red-tailed hawk perched in a tree, a mink or a weasel along the water's edge. There are cold days when ice freezes in the rings of the rod at every cast, still colder days when there is anchor ice on the bottom of the river, ice in every back eddy and even along the edge of the current, days when everything wants to freeze. I fish very carefully on those days, breaking the ice from the guides, trying to keep water away from the reel, trying to keep my fingers alive and sensitive because I know the fish will come gently if he comes at all. Many times I have had a fish take when everything was frozen, and at least twice I have had the leader broken before I could free the reel.

One does not fish long on such days and when a fish or two has been successfully hooked, beached, and released there is a real sense of triumph over difficulty. The cold has crept in at the wrists by now, the effort of freeing ice from the rod guides no longer seems worth while, fingers are cramped like claws and merely fumble on rod and line. It is time then to go back to warmth and shelter, time for the glow of overproof rum to slide along the arteries and unlock the crooked fingers. Such days are extreme and I fish them only occasionally, to prove to myself that fish can be persuaded to take a fly, no matter how cold the air and the water.

Coast winters have other days, mild and pleasant days with the thermometer in the high forties or even the fifties. I have fished July days that were far worse than these in the Canadian subarctic and in the polar winds of Iceland. On such days one fishes comfortably and confidently. The fish will move, if they are there, and the fly may be close enough to the surface for the boil of the rise to show. I have not proved it yet to my own satisfaction, but I am convinced that on days like these, even in midwinter, steelheads can be moved up to take a fly within half an inch of the surface or even on the surface.

This is what fishing is all about. Not just repeating over and over the things one knows can be done. Not just catching and killing. Not battling monsters of incredible strength and fury — the odds, after all, are nearly always with the fisherman. Not even in enduring harsh weather conditions. It is in developing and refining knowledge of the fish themselves and, with this understanding, finding ways of taking them that show them at their best. The steelhead is a complex and beautiful animal, as are all the trouts and salmons. The challenge is not: "Can you catch him?" It is rather: "Can you catch him in the way you want, with fullest respect for his qualities and a real testing of your own skills?" It is a fascinating challenge and one that never grows old.

10
First Among Equals
(1976)

Comparisons are, I suppose, inevitable. And though they are often damned as odious, insidious, and even meaningless — and though I dislike putting myself in the position of rating the fish I fish for — it is interesting to compare the performance and all-around virtues of the various fly-rod fish.

I propose to limit the discussion to the trouts, chars, and salmons, which is a pretty wide range in itself. And I intend to discuss them with deep appreciation for the splendid creatures they are, whether or not they have always chosen to co-operate with my own preferred ways of fishing for them.

Obviously a fisherman's individual preferences must tend to colour his views. I am, by strong preference, a stream fisherman and a fly fisherman. Yet when I think back over a lifetime of fishing I recall that I have had many good moments in lake fishing and some intensely exciting experiences in salt water and tidal water, and by no means all of them have been at the butt-end of a fly rod.

Like most salmon fishermen I know, I have served my time in search of "the big fish," meaning something in the next ten-pound range beyond anything I have caught before. With Atlantic salmon in my youth, I started at a properly modest ten pounds, went rapidly through the teens, including one of eighteen pounds, then to twenty-eight pounds. But my peak, still unsurpassed, was a fish of thirty-two pounds.

Most of these fish were taken on a prawn or a Devon Minnow, fished very deep and very slowly in deep, slow pools. Yet I remember the fish as being formidable. They hit hard and ran almost instantly from the strike. One rarely saw a fish before the strike unless it was a jumper, but the really productive lies were known almost to the inch. The anticipation and suspense of fishing down and into those lies was almost unbelievable, and the disappointment of fishing past them without a take was immense, even when the river had been unproductive for days.

Part of this emotion, no doubt, was youthful concern and involvement, but those were strong fish. They took charge and could strain tackle, rap careless knuckles with the reel handles, surge out of the water fearsomely and unexpectedly, and twist their gleaming bodies with a violence that no tackle should be expected to withstand.

I pursued this matter of big fish somewhat further with the chinook salmon of the Pacific Coast. Once I had learned how to catch them, thirty, forty, and fifty pounds came quickly enough, but sixty remained and still remains elusive. Those, too, were formidable fish, and I caught most of them by casting out a large spoon and fishing it slowly and deeply through a big pool.

But there were other times, on big flooding tides near the mouth of the river, when the schools were moving in. On a calm night one could see them, a formation of arrowhead ripples moving slowly on the gleaming surface of the water. The cast had to be just right — close enough, but not too close, perhaps ten feet ahead of the leading ripples. One arrowhead, sometimes several, would turn and follow, to take in a swirl and turn away, or perhaps be beaten to the spoon by some other fish, usually smaller, from the side. Sometimes one could see the shadowy shape of the fish under the ripples, and more than once I knew I had hooked my sixty-pounder, though I never brought him to boat.

It is difficult to ask much more of fish or fishing than this. If those salmon had been cruising in search of floating flies, it could not have been more exciting. Since that time I have caught a fair number of large chinooks on the fly, but I am not sure that I really think of them as fly-rod fish — not those monsters returning in the fall from four or five years in the ocean deeps. The smaller, feeding chinooks that can sometimes be found and taken near the surface in inside waters are another matter, and they must rate high in any fly-fisherman's scale of values.

And what of the rainbow? The rainbow is a true trout, an accredited member of the genus *salmo*, close relative of the brown trout, the cutthroat trout, and the Atlantic salmon. He may be a lake or stream resident, like his generic brothers; but as a steelhead he is also anadromous, an ocean-going, river-spawning creature with a life history almost exactly the same as that of the Atlantic salmon, and very close to those of several of the Pacific salmons except that he may live to spawn more than once. So it seems to me we must consider and compare the rainbow in three distinct forms: as steelhead, as lake resident, as river trout.

My first experience of rainbow trout was at the age of four or five, when my father planted some fingerlings in a small pond in the south of England. I can still remember the hatchery cans, much like milk cans, which we carried down to the water's edge; the small fish swimming rather nervously in them; the care with which we dipped water from the pond into the cans until temperatures equalized; and finally the great moment of release when the fish swam over the lips of the cans into the freedom of the pond. It was my father's third attempt, and like the others it failed. The fish disappeared entirely after a few months.

This disappearance was the subject of much discussion and theory and gave the little fish an added dimension of mystery.

Where did they go? The pond was spring-fed, about a quarter of an acre in area, and full of carp. It drained by ditches to the River Adur, and so to the sea. My father theorized that the little trout "got down in the mud and died." I suspect today that the summer temperatures were too much for them. But the theory I liked best was that they somehow found their way down the ditches and the river to the greater freedom of the sea.

It was another fourteen years or so before I caught my first rainbow, and that was a sea-run fish, a steelhead. In those days you could still get an argument among fishermen about whether a steelhead was a rainbow trout, but I believed firmly that it was. After all, that was where my father's fish had disappeared to, and here was the confirming evidence. I thought the fish beautiful and its performance splendid then, and I have had no reason to change the assessment since that time.

Theodore Gordon paid some of the handsomest tributes to the rainbow: "The rainbow is a great fish, playing like a salmon and having much the appearance of that fish. They usually lie in heavy, deep, swift water, and after taking the fly throw themselves into the air about three times and then rush downstream. . . . What gay fellows they are! They leap and fight until completely exhausted."

Gordon also admired the native brook trout of his eastern streams, but his final preference was for the introduced brown trout, especially what he called "the yellow variety." He wrote: "The yellow trout are real beauties. The vivid spots, golden belly, and general brightness of colouration can hardly be surpassed by any other fish." He believed they were hardy and prolific, growing faster than rainbow or brook trout, and he admired particularly their preference for insects and surface feeding. He even defended their fighting qualities, pointing out that they jump frequently in cold water and do not tire easily and that large, sulky fish can be "dangerous to handle."

I agree with all these opinions. A brown trout is very beautiful indeed (so also is a fine red-side, a well-coloured harvest cutthroat, or a fresh-run steelhead, to say nothing of the chars), and the brown probably offers a finer test of fly-rod skills than any other trout. He hasn't the dash and abandon of the rainbow or the steelhead or sea-runs of his own species, but in his own familiar stream surroundings he is more dangerous than the others, boring down, using his strength to best advantage, and making good use of any obstruction or shelter.

I have never practised the sport, now so popular in the mountain states, of trying to match fish size to fly size (for example, a twenty-two-inch fish on a No. 22 fly, or even a twenty-eight-inch fish on a No. 28 fly, which I believe has been achieved). I can imagine being in some degree successful in a rocky stream such as the Madison or the Gallatin, especially with rainbows. But on weedy streams such as Silver Creek, Armstrong Creek, or Odell Creek, I would guess that even an eighteen-inch brown on a No. 18 fly might well be too much for me.

I have caught brown trout from Chile and Argentina in the south to Iceland in the north, from Hampshire's Itchen and Perthshire's Isla to the Cowichan River on Vancouver Island, nearly always alongside other salmonoid species, and I think that browns must be rated at the top as all-around resident stream and river fish, not least because they learn fast and are almost invariably difficult to deceive by the time they are of respectable size. As lake fish they have quality because of their free surface feeding and their habit of cruising faithfully on a line instead of running away from their rises as rainbows do.

Even so, it is difficult not to think of the rainbow as the supreme lake fish, especially in the form that inhabits lakes of the interior of British Columbia, usually above the two-thousand-foot level. The Kamloops trout, as this form of rainbow is called, probably provides more lively and spectacular sport than any other lake resident. Most of the lakes of this region are slightly alkaline, rich in minerals, and highly productive. Many contained no fish until Kamloops trout were introduced from other lakes nearby. The result is a wonderful variety, from lakes with large numbers of small fish to lakes with a few very large fish.

I find the Kamloops trout disappointing in that he so often prefers to feed on the bottom of a lake and must be fished for down there. But whenever big fish choose to feed on or near the surface they can be extremely exciting. No other lake fish I have caught runs or jumps so splendidly.

Kamloops trout can also be superb in rivers. During the migrations of sockeye salmon fry and smolts from the Shuswap lakes, big fish move up to the mouths of Adams and Little Rivers to chase and slash at the small fish right on the surface. There are resident or semi-resident fish in the Thompson River that attain weights of twelve pounds or more and are almost indistinguishable from steelhead.

Whether the rainbows of the Stellako River, between François and Fraser Lakes on the upper reaches of the Fraser watershed, are properly classed as Kamloops, I am not sure. But I have taken Stellako rainbows of four and five pounds that were rising to surface flies as delicately and daintily as any brown trout, with the same almost imperceptible swirls on glassy-smooth glides. The movement of these fish from the strike is explosive and fully confirms Theodore Gordon's observations. The fish are usually beached, if at all, in the next pool downstream. Even so, if I wanted to try for, say, a twenty-four-inch trout on a No. 24 fly, I think the Stellako is the river I would choose for the attempt.

The cutthroat is a quiet, almost humble fish. Like the rainbow, he is a native of the Pacific slope of North America. Unlike both rainbow and brown, he is not an easy fish to raise artificially, or to transplant, so he has nothing approaching the world-wide fame of the two others. The interior forms of the cutthroat are often highly coloured and very beautiful.

I never saw any of the great Pyramid Lake cutthroats that reached weights of more than forty pounds, but they must have been impressive, and it is good to know that stocks are rebuilding after the abuses of the past.

But the cutthroats I am thinking of are the coastal, sea-running cutthroats of Oregon, Washington, and British Columbia. These are splendid tidal-water fish, found at times in and near the estuary of almost every northwest creek and stream and river. In tidal water they are usually bright and silvery, sometimes almost indistinguishable from steelhead though somewhat smaller. The big maturing fish that come into the rivers in late summer and early fall may be silvery at first but soon take on the bright-red slash and the black-spotted gold of their breeding colours.

Years ago these harvest cutthroats — or yellow-bellies, as the irreverent like to call them — gave me some of the finest trout fishing of my life as they came, every year, into the Campbell River. By mid-July they were in the great wide gliding pool that fanned out from the mouth of the canyon before the hydroelectric company built the powerhouse there and destroyed it. It was a bit of a walk in to the pool, so almost no one came there. The wading was deep and awkward, and the fish were mostly far out and hard to reach, in smooth, gently moving water five or six feet deep. A ripple from one's waders or a bad cast disturbed them and put them down for an hour or more, so one crept out over the rocky bottom and stood motionless for several minutes before even starting to cast for them.

They would come to a floating fly, but it was a bad method because one could reach them only by casting almost straight across and slightly downstream. Drag developed very quickly and swam the fly all across the tail of the pool. A rise and the resulting strike produced a massive disturbance on the smooth surface of the pool. Instead of a floating fly, I used the standard greased-line technique for those fish — a floating line, a 3x leader, and a No. 6 or 8 fly sparsely dressed on a slender low-water hook, and fished just under the surface film with as little movement as possible.

Most of the fishing was on calm, sunny afternoons, which was pleasant for me, and gave the fish some advantage. When a fish rose I tightened very gently from the side, letting the pull come around the belly of the line, and tried to lead him quietly away from where his friends were lying before starting to fight him. On a good afternoon when everything went well I would net half a dozen fish or more, averaging over three pounds apiece. Very few of those fish were smaller than two and a half pounds.

Given the right conditions it is possible to do the same thing with summer-run steelhead or Atlantic salmon, and since either of these species is likely to average two, three, or four times the weight of those cutthroats, it would seem that they must rate higher. I am not sure they really do. There was something about those cutthroats, something essentially trout-like in their behaviour and appearance, that is lacking in the true sea-going anadromous fish.

I can make the comparison because I found in the end that summer steelhead were lying in the same pool, at the same time, farther down toward the tail. I caught them up to sixteen pounds by the same method with the same flies — but I learned to go up to 1x or 2x leaders. They were fine fish that broke me or took me out of the pool as often as not, but somehow they were less important — interlopers in the pool that really belonged to the cutthroats at that particular time.

Having expressed this base ingratitude, let me now turn to praise of the summer steelhead. He is a fish that should never be taken on anything but a fly and a floating line, unless the water is very high or the weather very cold, and it is always worth giving him a chance at a floating fly. In many streams, especially in mid- and late summer, I have found that a floating fly produces more action than anything else. And to watch a ten- or twelve-pound fish take down a big hair-wing fly in a slow motion head-and-tail rise is a pretty high experience.

Summer steelhead jump well and run hard and far and fast,

though they can vary greatly in performance, usually in proportion to the length of time they have been in from salt water. Perhaps they really do rate number one. There have been times when beating heart, trembling hands, and jolting adrenalin have fully persuaded me on the point. Yet there is always some other fish at some other time and place to insist that the question remain open.

Twice in recent seasons I have been lucky enough to fish for Atlantic salmon on a small river in northern Iceland. There were plenty of salmon, taking freely at some time during most days, and I learned by trial and error that a small fly fished right up in the surface film, with a little added speed and action, was what they wanted. Pattern did not matter much; size was everything. I settled for a No. 10 Blue Charm and securely hooked almost every fish I rose. On a No. 8 I lost more fish through the hook coming away, and on a No. 6 far too many.

Most of the fish ran from nine to fourteen pounds, though a few were larger. Every rise was visible, and most were preceded by a visible follow. When the salmon were really on the take, most of them intercepted the fly on the swing, moving slightly upstream with it. In less eager moods they followed the fly around and took from directly below it, though they still seemed to move up as they took. In either case I made no movement of the rod until I felt the delayed pull, and then it was a smooth tightening rather than a strike. It was so exciting and satisfying that I didn't really want to waste time fighting the fish; I wanted to get the fly free, make another cast, watch and feel another rise, tighten on another fish.

My first impression of those fish was that they didn't fight very hard. Most of them in the nine-to-twelve-pound range ran once some twenty or thirty yards into the backing and jumped or surged out of the water once or twice. I would recover most of the line, and the fish would run again, no more than a few feet beyond the backing splice. Then he would be in close, and I would think I could beach him and get back to my fishing. I was fooled every time until I realized they simply did not kill themselves with long runs and violent jumping as a steelhead does. They saved their strength for the most dangerous time, when they were on a short line.

Almost invariably there was a good five minutes of strain and sweat between the time I first brought a fish in close and the time I could net or beach him. Beaching was never easy, because it was almost impossible to get the fish's head well up. A sudden run of twenty or thirty yards, a violent jump, or a flip-over was always possible. I netted most of them with a good, big net, but I was never able to bring a fish over the net on his side and lift in the usual way. They had to be led head first into the net until it was too late for them to turn. One hit the net so hard that he took the head right off the handle.

What impressed me was that these fish were more difficult and more exciting to land than the wildest of steelhead simply because of that unpredictable reserve of strength when they were close in. I could not recall anything else quite like it, except perhaps the late strength and resistance of a chinook five times their size.

So there is no firm and fast conclusion. The fish that rises the way you want him to, that thrills you and makes things difficult for you, that pleases you with his strength and grace in action and delights you at the last with his beauty, is always number one. But he comes in many species, in many ways, in different times and places, and there is always something special about him that is not shared by the others.

11
Grandmother, What Sharp Teeth You Have!
(1960)

It is the hope of taking a gigantic fish that keeps one fishing for northern pike. They have the quality of inspiring this hope — somewhere, under lily pads, over weed beds, in this dark hole or that river eddy, lurks leviathan. If only one keeps on trying, sooner or later the pike will come, if not on this cast, then on the next one, a sudden, broad-backed shadow, cold-eyed, swift, hungry, ferocious — and every inch a lady.

This last may come as a shock to some fishermen who have spent days and weeks on northern lakes and rivers in search of that one big fish, but the fact of the matter is that when it comes, it will not be the great-granddaddy of all northerns but the great-grandma. For the biggest pike are females; the males rarely exceed nine to ten pounds in weight, while the ladies of the species will run five or more times that weight. But, male or female, they are not difficult to catch when in a taking mood. Almost any moving lure, from pork rinds to plugs, from flies to spoons and spinners, will interest them. One fisherman was highly successful with a huge contrivance carved out of balsa wood, clothed in muskrat skin, and provided with little paddle-wheels to make a surface disturbance. The clumsiest cast will not frighten pike — in fact, a spoon landing with a good heavy splash usually serves to draw attention to itself.

Thus the hope of catching a really big northern is not, on the whole, unreasonable. *Esox lucius*, the common pike, is a fish circumpolar in distribution in the northern hemisphere. The North American record is a fish of forty-six pounds, two ounces, caught in 1940 by Peter Dubuc in Sacandaga Reservoir, New York. In 1941, a net-caught fish from Cree Lake in northern Saskatchewan was weighed at fifty-five pounds, and there are plenty of other well-founded stories which make it clear that the present rod-caught records will not last.

Northern pike are bold fish. They like the sun as few other fish do. Often even a large female can be easily seen, lurking in weeds or openly basking in sandy bays. From above the surface of the water, her dark back open to the light, she seems not to hide at all. Down at her own level she is well-hidden by the motionless pose of her body, perfectly maintained by tiny fin movements, and by the mottling of her sides that is as deceptive as underwater shadows. A large pike rarely strikes from a distance of more than ten or twelve feet, and she does so with a sharp rush of speed that is calculated to bring her exactly alongside the prey — bird, mammal, or fish, perhaps her own grandchildren. There she is likely to pause, then turn abruptly to seize the chosen creature squarely across the body. The long, sharp teeth of the lower jaw hold her prey securely, and she waits out its struggles. When her victim finally is still, the pike turns it in her jaws, and the raking teeth of the tongue and vomer and palatines force its head foremost into her gullet.

A good, big northern, grown far beyond her fellows, could be almost anywhere. Young northerns have the fastest growth of almost any freshwater fish. A big female produces more than a hundred thousand eggs, depositing them in weedy bays and marshes as soon as the ice breaks up to let the fish run in from the lake. The eggs hatch in seven to fifteen days, depending on the water temperature, and in another three or four days when the yolk sac is absorbed they become tiny, free-swimming creatures about a quarter-inch long. At that stage they are prey, not predators; not more than one or two in a thousand are likely to survive the two or three weeks they usually spend in the spawning marshes before migrating to the lake. The few that do survive the first year will be about ten inches long in most United States waters, eighteen inches long by the end of another year, thirty-three inches at six years, forty inches at nine years, from forty to forty-seven inches at ten to thirteen. In Lac La Ronge in Canada, on the 55th parallel, a forty-inch fish is likely to be fifteen years old, a forty-six-inch fish perhaps nineteen or twenty.

Northerns grow fastest in the more southerly part of their range, so the logical place to look for a really big one should be south of the Canadian border. But southern waters are hard-fished, and though a female northern may grow to twenty pounds and forty inches in ten years there, she is not very likely to get a chance to do so. The most promising and exciting northern pike waters on the continent are in the far north, across northern Manitoba and Saskatchewan, where the lakes are innumerable. Some of them, like La Ronge, Reindeer, Cree, Wollaston, and Athabasca are very large. Northerns grow more slowly in these latitudes but, like trout, they live much longer as a result. Because few people have been fishing for them, the far-northern northerns are still swimming about in their old age, occasionally scaring the wits out of anglers, and from time to time making trophies of themselves.

Henry Weitzel of Cree Lake knows of a fish in a bay off Sand Island that is as big as a small canoe. Henry and his friend Martin Engman have lived on Cree for many years, and I would venture to say they are growing old there, except that they are ageless. Both are very good men to fish with. Henry knows his lake and his northerns as though they were kin. Cree is certainly one of the best of the northern pike lakes. It is big (350 square miles) and beautiful, with more than five hundred islands, and beaches of pinkish sand. Its waters are clear and cold. Temperatures in the upper thirty or forty feet rarely exceed 60° Fahrenheit. Thirty-pound northerns are caught there every year — a fish of thirty-six and a quarter pounds was taken in 1958, and there is not the slightest doubt that bigger fish are in the lake than ever have been taken out of it by rod and line. I spent a cold and stormy midsummer day with Henry on Cree Lake, looking for a twenty- or thirty-pound northern.

"They're all coming out of the shallow water about now," Henry said as we started. "You've got to look for them off the rocky points — in the lake trout spots. Better on a sunny day."

He was right about the rocky points. The fish were there, dozens of them — six-, eight-, ten-, twelve-pounders, a few larger than that, but not twenty pounds.

Henry released them and scolded them: "Go away. We don't want you. You should have let your grandmother have it, but you're too stupid for that. You had to go and jump in ahead of her."

Occasionally he decided to keep one of the larger ones to feed his sled dogs. Once he picked up a junior-sized baseball bat, waited until I had brought the fish alongside, and whacked her squarely on the head. Instantly she took off in a twisting jump like a crazy creature. "That caught your brain box," Henry would tell her. "Teach you some sense."

The northerns have the same disadvantages as fighting fish that they have as predators. They are fast, ferociously armed, bold, and voracious. But, like many very fast predators, they can't sustain speed (it has been measured at twenty miles per hour, compared to twenty-three for the trout and twenty-five for the salmon) for more than a very short distance, usually not for more than twenty to thirty feet. In this the pike is like the mountain lion, whose speed over a hundred feet in four or five long leaps is great enough to catch anything that runs, but whose narrow chest and shallow lungs check him sharply at the end of his first rush. Even so, the northerns have plenty of speed to overtake minnow, perch, and bass that make only eight to twelve miles per hour. A more serious limitation is their stiffness in the water and their need to capture prey broadside. They cannot turn and twist in pursuit as trout and salmon do. A big northern female makes her rush very straight, hovers over her victim, and only then turns her still body to take the victim in her jaws.

When large pike are hooked, they do not often jump clear of the water but may lunge out in brief, spectacular head-shaking flurries. Big fish often come in towards the boat and lie like logs until they are lifted; they have all their strength then, and can take off so suddenly that a clumsy hand on the reel can mean disaster. The horny jaws and gill-covers have many soft spots, and hooks that seem securely set can readily tear away. The northern has ways of making the fisherman feel he has earned his catch, and not the least of these is the big fish habit of following a cast in to the boat and fading back at the last moment without striking.

I suppose we caught thirty or forty northerns just south of the Widdess Bay in Cree Lake during the morning. After lunch we rode down the wind from Widdess Bay to Frog Bay, the motor labouring up the swells and racing with the crests, the twenty-foot canoe like a live thing under us. In Frog Bay it was much the same story as before, except that half the fish in the lake seemed to be congregated off one rock-piled point. Not twenty-pounders, but it wasn't that they weren't in the lake — twenty-pounders had been caught on the two previous days.

Wollaston and Reindeer, larger lakes north of Cree, and Athabasca, six times as large, lying across the 59th parallel, have big northerns, as do lesser lakes like Riou, Hatchet, Black, Careen, and many others. The Churchill River, flowing through Saskatchewan and Manitoba into Hudson Bay, is a series of lakes joined by rapids through much of its length, which produces northerns of twenty to thirty pounds each season.

This northern country, with its rocky hills, its sand bluffs and gravelly eskers, the small-growth jackpine and birch, the carpets of reindeer moss, and the subarctic feeling of its short summer, is new and exciting. It has an abundance of lake trout, walleyes, and Arctic grayling, as well as its great northerns, and no fisherman is likely to come back disappointed from it, least of all the pike enthusiast. Over the next twenty to thirty years some records are going to be broken up there, and while this may not be all-important to most fishermen, it is always more exciting to know that really big fish are around.

How big the biggest may be, in North American or any other waters, is obscured by the wealth of legend and folklore and fish stories that the pike has inspired in Europe since medieval days — the Mannheim pike, for example, that was supposed to be nineteen feet long and two hundred and sixty-seven years old. As the British naturalist Francis Trevelyan Buckland put it, the pike has produced more lies than have been told about any other fish in the world. But since the European pike is biologically identical to our northerns, it is well to examine some authentic old world records. A fish of seventy-two pounds, caught in Loch Ken, Scotland, in the nineteenth century, has been given some support by experts, through measurements of the preserved skull. The accepted rod-caught record is, so far as I know, a fifty-three-pounder from Lough Conn, Ireland, by Mr. John Garvin in 1920. But the greatest fully confirmed weight is a fish of 58.42 pounds, fifty-nine inches long, caught in Sweden on a set line. That blank of the western angler's map of the world, Russia, has pike, and presumably some very large ones. Perhaps one day a record will be claimed from Siberia or Mongolia or Turkestan. The pike there is the same fish, and no doubt fishermen are out after them on great rivers like the Don and the Volga.

The pike has never given rise to legends in America as it has in Europe. I can think of several good reasons for this. On the North American continent the pike has four close relatives — the muskellunge and the three pickerels: grass, chain, and redfin. The muskellunge, as a somewhat larger and similar fish, has more readily caught the imagination. The northern has had all the other wonders of a new continent to compete with. And I suppose we are a less credulous people, at least about natural phenomena, than were our European ancestors of a few generations back.

On the whole, the reputation of the northern has been rather overshadowed by that of the muskellunge, and that is a pity. The muskie may, it is true, grow into a bigger fish, but under modern angling pressure rarely does so. In general habits and performance, there is no great difference between the muskie and the northern. A few anglers believe that the muskie is a stronger fighter and more likely to jump when hooked, but most seem agreed that there is little to choose between the two.

Whatever his rating as a game fish, the northern provides a lot of satisfying sport over a wide area, often in waters not especially suitable for more highly regarded fish. It is true that in some waters the pike prevents the introduction of more desirable species, and in others both preys upon and competes with species such as whitefish and lake trout that may well be more important, both commercially and as game fish. In a few areas, notably the Saskatchewan and Athabasca deltas, it has been estimated that the pike destroys a million and a half waterfowl a year. But if the big pike, the voracious lady of the lake, is a villain, she is at least a native villain, and quite a romantic one, honourably established in a pattern of underwater life that has always served North Americans well. The wise course is to recognize her many virtues and put them to good use.

12
A Westerner Looks at the Beaverkill
(1959)

One comes to it as a pilgrim to a shrine, a sacred, shining place, made so by the deeds of the great and passing feet of many generations. And it is just that.

There is sense and no sense in picking Opening Day for the pilgrimage. The noble hills are brown except for scattered conifers; the hardwood trees are brown and leafless and last year's leaf fall is brown under them where it is not whitened by the passing snow flurries. The air is cold, the trout lie quiet in the deep pools, when they are not down in the still deeper pools of the East Branch. It is an unlikely time for good trout fishing.

Yet it is the gathering time for pilgrims old and new, and the time of all times for a new pilgrim to learn the devotion that the little river inspires and to sense its traditions. For the Beaverkill, the Neversink, Battenkill, Willowemoc, and a few others, are surely the nursing waters of American fly fishing.

On Friday night, with the frost tight in the ground outside, the bar at Doug Bury's Antrim Lodge was crowded with fishermen greeting each other for the first time in six months or a year. It was too cold, they agreed, too early, and there were far too many people around. "Fishing tomorrow, Jim?" "Naw. Just go out and look it over." "Fishing tomorrow,

Frank?" "Drive up and down a bit, maybe, talk to a few guys. Might put up a rod." Never have so many fishermen driven so far to protest their intention of not fishing. Yet I knew just how they felt. Tomorrow was Opening Day. There would be few lines that did not touch river water, however briefly, and those would mostly be spares.

Bill Naden of the Brooklyn Fly Fishers Club arrived in good time next morning, and he and Ed Zern showed me the river — downstream first, from the Junction Pool at the mouth of the Willowemoc clear to the East Branch of the Delaware. Many miles of beautiful water on a big, bold stream, nearly all of it readily accessible from the road. On Opening Day, little more than a hundred miles from Manhattan, I had supposed it would be crowded with fishermen. True, they crowded a few favoured spots, lined both banks five or ten feet apart at the Junction Pool and at a slab rock pool by a bridge; but there were miles of water where a man would have found no more competition than on any British Columbia stream within easy reach of a paved road. And fly fishermen, may the Lord sharpen their hooks and guide their wrists, seemed as numerous as spinners.

We drove back, stopping now and then to watch and talk and uncork a bottle, and came to the upper stream above the Willowemoc junction. The shades and living spirits of the great were now with us, their names on our lips: Hewitt with fourteen-foot leaders, La Branche with nine-foot or less — "That's why he doesn't catch any fish," said the shade of Hewitt, careless of fact. Sparse Grey Hackle, who will take many more fish from the river on No. 18 and 20 flies; gentle Jack Atherton who, unhappily, will not. And Theodore Gordon: "For at least a hundred years," he wrote over fifty years ago, "the Valley of the Beaverkill has been celebrated for its beauty and the river for its trout."

Gordon could well write the same words today if he were living. Clear and cold and lovely, the upper stream sparkles down in cascades and runs and trout-loved pools from its spring-fed source in Balsam Lake. We saw fishermen lined again close by Ogden Pleissner's covered bridge. We saw miles of posted water broken by shorter stretches of free water. I thought of Gordon again: "I fear that in a few years very little water will be free to the public . . . for real sport, give us free water, where the trout are critical, hard to please and highly valued when caught." Gordon believed that to take large fish when they were shy was the acme of sport.

We wet our lines in the posted waters, working down the Brooklyn Club's Home Pool, through Twin Rocks and on the lip of the dam, where we swung big streamers back and forth in the foaming overflow, while a brief blizzard hid even the nearest bank from sight. We turned back fishless to the big fire in the clubhouse, but tradition had been served and well served, and we were happy as men can be.

Western eyes and a western heart could not ask for more than this. It has all been going on for a very long time. The fathers of the sons and the grandfathers of the grandsons fished here on earlier Opening Days, grew up with the clubs or on the free water. Halford and Skues and Pulman, Walton and Cotton and Berners made spiritual pilgrimage across the ocean to the Beaverkill. Gordon, Hewitt, LaBranche, Atherton, the Darbees, the Dettes and many others caught the spirit, interpreted it anew, and sent it out across the vast continent. Today it reaches the many millions who seek respite and reward in the fly streams from Atlantic to Pacific, from Mexico to the Arctic Ocean.

I would go again to the Beaverkill when the rhododendrons have opened their buds on the slope above the Home Pool, when the trees are in full leaf and the air hums with warmth and life and the river is low. I would hope for a hatch and a few great trout rising steadily here and there, "critical and hard to please." But if there were none, it could scarcely matter. The river would be there and its banks — and all about me the shades of great fly fishers. They would know I searched faithfully, and in an honourable tradition.

13
Chilean Trout Fishing
(1954)

Most of us have heard wonderful reports of trout fishing, only to follow them up and find that they were somewhat larger than life-sized. I spent years in British Columbia chasing up rumours of lakes and streams where the trout "averaged three pounds and came readily to the fly," and was nearly always disappointed until I learned to divide rumours by two and remember that this still meant first-class trout fishing. Most fishermen have heard wonderful reports of trout fishing in Chile. The rumours were of five- and ten-pounders; in reality, the trout averaged three pounds or better, which is fabulous trout fishing in any land or language if the right conditions go with it. In Chile they do.

This is not to suggest that one cannot find really big trout in Chile. They are there, rainbows and browns of five, ten, fifteen, and even twenty pounds. But in most parts these are the exceptional and memorable, to be reasonably hoped for but not confidently expected — which is exactly as it should be in trout fishing. I know of only one place in Chile where a five-pounder is a small fish and trout of over ten pounds are commonplace, and that is Maule Lake, about two hundred miles south of Santiago and seven thousand feet up in the alpine desert of the Andes. The first fish I took from Maule weighed nine and a half pounds. Later, fishing with a fly, although I had been warned that the fish would take only spoons and plugs, I caught six fish between five and ten pounds in two hours and lost several others.

Maule is a magnificent lake, full of feed and wonderfully productive. Its fish are all rainbows; they take freely and fight well. The surrounding country, with snowbanks and pumice cliffs and desert flowers, narrow green valleys and mountain peaks, is grandly exciting. But Maule is in no way representative of Chilean fishing. The real Chilean fishing is in the south, in the swift and lovely rivers that go down among rain forest and farmlands. It is a world of little Swiss and German hotels adapted to Chilean ways; of cowboys and ox carts along dusty roads; of wild fuchsia and white ulmo in splendid blossom; of bamboo shoots and scarlet-flowered vines and great hardwood trees and many birds; of skilful boatmen and powerful rapids; of a hundred rivers that produce rainbow and brown trout in abundance. Chilean trout fishing is an adventure in travel through beautiful and wholly fascinating country, among people who are at once intensely courteous and genuinely friendly.

Temuco, an attractive city about 450 miles south of Santiago, is really the beginning of the fishing country. It is true that there is some fine fishing in rivers north of there, notably the Laja, the Nuble and the Diguillin, which can be reached from Yungay or Chillan, but to get the best out of it one needs some local help and advice. At Temuco one can arrange for boats, guides, and transportation to two excellent rivers nearby, the Cautin and the Quepe. Between Temuco and Puerto Montt, three hundred miles farther south, where the roads and railroad end, the country is full of lakes and rivers and there are at least a dozen excellent fishing centres; Villarica, Pucon, Los Lagos, Llifen, Puerto Nuevo, Osorno, Puyehue, Puerto Varas, and Puerto Montt itself, to name some of them. All have adequate hotels, and some are first-class hotels, which gladly assist visiting fishermen by providing guides, boats, fishing lunches, and transportation to and from the rivers, or else accurate information as to where these necessaries may be found.

The best of Chilean trout fishing is in the rivers, and with a few exceptions, such as the Golgol and the Petrohue, one travels the rivers by boat. This is a very comfortable and profitable way of doing it. Most of the boatmen are expert at running the rivers, some of which are very difficult and exciting, and a good proportion of them are skilful fishermen as well. Many anglers simply sit in the boat with flies or lures dragging astern and let the boatman swing them back and forth across the current and through the favourite places. This produces good results, especially in the early-season months of November and December, when rivers are high with the spring run-off.

It is possible, however, and even more profitable in most streams, to do nearly all one's fishing by wading and casting. Once a boatman gets the idea that this is what his fisherman wants to do, he happily transports him to the best places for wading and puts him out to go to work. One gains the important advantage of fishing much good water that the boat fishermen necessarily pass over all too quickly, as well as the greater pleasure of fishing properly.

I remember now the Tolten, below Villarica Lake, with its high green banks, a big river beautifully broken and full of fish; the Liucura, river of furious white water and magnificent boatmen, with big rainbows lying in difficult places; the Lower Trancura, bigger and fiercer still, beginning with the superb Martinez Pool, one of the truly great pools of the trout fisherman's world, and ending with the big fish on the bar where the river breaks over into the head of Villarica Lake. And on all three rivers the recurring sight of the long, smooth white slopes of the volcano Villarica, with a trail of smoke from its windy peak. I remember the lively, dancing Calcurrupe, hurrying between Lago Maihue and Lago Ranco, with four-pounders lurking in its pools and still larger fish at either end; I remember Cautin and Quepe, Enco and Fui, and the two big boat rivers, Rio Bueno and Rio San Pedro, where gigantic brown trout of fifteen and twenty pounds are caught every year.

One starts out early enough, but not too early in the morning — 7:30 or 8:00. There is usually a crowded drive upstream, in a taxi or truck, to the place where the boats go into the water — one's own boat, a friend's, perhaps another boat or two that are making the same run that day, though I never found even the favourite rivers in any way crowded. There is some sort of mutual agreement about where and when to meet for lunch, and the boats push off.

I usually came quite quickly to some good broken water where I wanted to get out on my feet and cast a fly, and from then on the day would take hold. I don't remember a bad day, and only one or two when fish were really hard to find. The legal limit in Chile is thirty centimetres, or twelve inches, but I held pretty well to a sixteen-inch limit except for a few smaller fish killed each morning to be eaten at lunch time.

On one fairly typical day's run I jotted down the time and place of every fish incident as it happened, something I have never attempted before, or since, and found that in eight hours' fishing I had recorded forty incidents, including fish lost, returned, or missed on the strike. At the end of the day I had twelve fish in the boat, the largest a brown trout of four and a half pounds, three fish of over three pounds, and eight others weighing between two and three pounds each. It was a satisfying and busy day that kept us on the water from 9:00 A.M. till 12:30 and from 4:00 P.M. till 8:30 with a generous three-hour break for lunch.

Lunch hour is not treated lightly on Chilean streams. The boatmen insist that it is the "bad hour" as far as fish are concerned, and really not one hour at all, but two at least, and preferably three. Every river has known and favourite lunch places, usually grassy, tree-shaded spots a few feet from the water's edge. The boats arrive within minutes of each other and within minutes of noon, almost without fail. One boatman goes to work to clean and fillet half a dozen trout; another builds a fire; a third sets the wine to cool in the river and packs the rest of the stuff up from the boats.

The fishermen do nothing at all. The fire, a big one, burns up fiercely and dies back to hardwood coals. The trout sizzle in black butter. The wine is cool, and someone brings it up from the river, a full bottle to each boatman and fisherman. There is a salad — lettuce and tomato, oil and vinegar, and hot green peppers. Several pounds of lamb or beef, impaled on a bamboo stake, roasts slowly over the coals. Overhead the wind rustles in the leaves and stirs the shining ulmo flowers. A great scarlet-headed woodpecker calls or an arrow-swift flight of green parakeets lights briefly and darts on again.

If Walton himself had looked down from heaven and tempted me, I would not have traded those Chilean lunches for a dozen record trout. The food was always good, and the wine was even better.

We seldom left the Chilean rivers much before dark, and then there was often a long drive back to the hotel. But the standard dinner hour, around ten or ten-thirty, is ideal for fishermen, and there was usually time to clean up and spend a few minutes in the bar.

I thought Llifen, at the head of Lago Ranco, just about the best fishing centre in Chile. From it one can reach four good streams flowing into Lago Ranco and four others flowing into Lago Maihue, as well as Maihue Lake and the Calcurrupe River.

There is good fishing at Llifen right through the season, from November till April. But there, as nearly everywhere in Chile, the best months are November and December, March and the first half of April. November and December would be like May and June in the northern hemisphere, months of the spring run-off when big fish are on the move, and some monsters are taken every year at that time on plugs and spoons. In March and April, our September and October, the streams are cooling again after the hot summer months, but they are low and accessible and the fly fisherman has his best chance. There is plenty of good fishing everywhere in January and February, but the streams are generally too warm then and really big fish are hard to find.

Though spoons and plugs are widely used in all the rivers of Chile, the fish come well to the fly anywhere, and no self-respecting fly fisherman need bother with anything else. Chilean streams hold all the same insects as northern trout streams — stoneflies, May flies and sedges, cranefly, deerfly and midge larvae. The trout take them, at times very well; but there is also an abundance of crayfish in the streams, and the larger fish are feeding on them more often than not. For this reason big, dark flies are most commonly used, though I found I was able to catch fish on a wide range of sizes and patterns. At times, especially on May fly hatches in the Martinez Pool and in one or two small side streams, the dry fly was very good indeed.

It is worth recording that Chile has no mosquitoes, no blackflies and as far as I could see, no hornets or yellow-jackets. There is a great blundering horsefly called the *colihuacho*, a smaller deerfly called the *tabano*, and a little evening midge called the *polco* or *petro*; I saw all these at their worst, but they were never more than a minor inconvenience. There are no poisonous snakes and only one poisonous tree, the *littre*, which I did not find south of central Chile.

On the positive side, Chile has a wealth of beautiful bird life along every stream, and the summer air is full of bird sound. The little parakeets are always exciting to see; the lapwing (*queltegue*) and the ibis (*bandurria*) are so much a part of Chile that I could not hear them anywhere again without thinking myself back there. Scarlet-breasted torrent-ducks play among the rocks of the wildest rapids. Black cormorants flight along the streams. There are many hawks, two lovely white egrets, a little night heron, occasionally flamingoes. Mocking birds, pigeons, doves, woodpeckers, and flickers are in the trees along the river banks. There are southern versions of our own familiar pintail and wigeon and teal as well as a great spectacled bronze-wing duck and several unfamiliar geese.

When people ask me, "How good is Chilean trout fishing?" I hesitate to answer that it is the best in the world. Although I haven't fished everywhere, I have enjoyed trout fishing mightily in many places. But when I remember wading out into the first rapids of the Petrohue or Rio Bueno and watching three- and four-pound rainbows shouldering up through white water to tear at my fly; when I think of the volcano Osorno, rising mightily from the plain of Villarica, shadowing its lovely lake; of El Tronador, Calbuco, Sierra Velluda, and the other splendid mountains; when I remember the Martinez Pool and the Trancura Bar — I wonder what more could a trout fisherman ask?

14
Ever Fish for Sebagos?
(1954)

Like most North American freshwater fishermen, I have dreamed at one time or another of fishing for sebagos, the landlocked Atlantic salmon of Maine and a few favoured eastern Canadian watersheds. Angling literature is full of references to their gameness and beauty, and a hundred fishermen have told me about them in exciting detail. But it never occurred to me that I should meet with them first south of the equator. In fact I did not know, until I was getting ready to go down on a fishing expedition to Chile and Argentina, that the sebagos had been successfully established there.

Efforts to introduce salmon and trout to the southern hemisphere, where none is native, have seldom been easily successful. It took many years of experiment to establish two trouts and one species of Pacific salmon in New Zealand; and there has been no success at all in establishing Pacific or Atlantic salmon, except this landlocked variety, in South America, in spite of repeated attempts by both Chile and Argentina. The sebago salmon is one of the most specialized and narrowly distributed game fish of the world. Even in Maine it occurs naturally in no more than half a dozen watersheds, where some obscure condition has worked to separate its habits from those of sea-running Atlantic salmon.

In Chile I had been asked to watch closely for any signs of success from the many recorded plantings of Pacific salmon. I did so, faithfully, but though I found rainbow and brown trout in wonderful abundance, I saw no sign of Pacific salmon, nor did I meet anyone else who had come across them. As I worked southward and over toward Argentina I warned myself not to be too sure of finding sebagos.

But they were there, as I quickly learned from Señor de Plaza, a courteous gentleman who is in charge of the hatchery at San Carlos de Bariloche. Bariloche is a pretty little town, not altogether unlike our Jasper, on the shore of Argentina's magnificent mountain lake, Nahuel Huapi. Like Jasper, Bariloche is the centre of a great national park, and it is surrounded by the superb peaks of the Cordillera as Jasper is by the Rockies.

The first shipment of sebago salmon eggs reached the hatchery at Bariloche — from another hatchery in Germany, not from Maine — in 1904. The young fish from this and subsequent hatchings were released in Nahuel Huapi Lake, which is over sixty miles long, in Traful, Meliquina, and other lakes that are also tributary to the Limay River, as well as in the Manso chain and in Lago Lacar, both of which drain to the Chilean side of the Andes. They established themselves quickly and by the 1920s the Traful River was yielding an average weight of eleven pounds, which compares to a North American average of about five pounds. In 1936, a sebago of thirty-six pounds was caught by trolling in Nahuel Huapi, and the North American record, a fish of twenty-two and a half pounds, was caught in Sebago Lake in 1907.

This is a remarkable adjustment. The sebago lakes of Maine are comparatively small, at low elevation and near salt water. The Argentine sebago lakes are large and deep, at an elevation of over two thousand feet, and nearly a thousand miles from salt water. The sebagos there must make their growth chiefly by feeding on crayfish, while the fish of Maine make their big growth by feeding on the smelt runs. Argentina's April is Maine's October and the fish must adjust their family arrangements accordingly. And no doubt there are many less obvious differences between the watersheds of the Appalachians and those of the Andes, all of which the fish seem to have accepted easily, and without any major change of habit or appearance. There is no real doubt that the "landlocked" habit of the Maine fish has persisted in South America. As in Maine, they grow to mature size by feeding in the deep lakes, then run up to the tributary streams or drop back to the outlet streams to spawn. Those of the Limay watershed are barred from salt water by the warm, muddy flow of the Rio Negro, which receives the Limay three or four hundred miles from the sea. Some of the other watersheds are physically open to the sea, but their fish have the same habits and appearance, and everything suggests that they remain voluntarily "landlocked."

I learned all this fairly quickly from Señor de Plaza and from other friends who knew the locality, and I began to wonder about catching the fish. A certain number, it appears, are caught by trolling in the deep lakes right through the season, and the fly fisherman can sometimes find them on the shallows or off the stream mouths. But there was nothing in Argentina to compare with the smelt runs of the Maine lakes, which bring feeding fish to the surface and up into the streams and make streamer flies so effective at times. The Chilean *pejerrey*, a small lake "herring" related to the silversides, is undoubtedly taken freely in its fry and fingerling stages, but seems to have no seasonal mass movements that anglers depend upon.

The main food of the Argentine sebagos is unquestionably the crayfish that are abundant everywhere, and the fly fisherman's real opportunity comes only when the fish drop back to the outlet streams in February and March — months which correspond to August and September in the northern hemisphere. I have a written "drop back" advisedly, although the movement is definitely a spawning run and the fish work up from the lakes as well as down. But the entering mountain streams are small, and by far the best fishing is in the big outlet rivers such as the Limay and the Traful.

Traful Lake is a very beautiful, stormy sheet of water twenty miles long and two miles wide, which is reached by road from Bariloche. Winds from the mountain valleys cross and chase each other all over the lake's surface, and under a blue sky it can take on a dozen different shades of reflected blue. Water comes down to it in little streams from the high Andes, among bamboo thickets and hardwood stands and pine forest. It is drained by the Traful, a good-sized stream that drops swiftly away to range country. And it was on the Traful in mid-February of 1952, that I found my first sebagos.

I was lucky enough to be shown the river and the fishing by Dr. Edlef Hosmann, secretary of the Norysur Fishing Club of Buenos Aires. Tito Hosmann has known the Traful country for twenty or thirty years and is certainly one of the best fly fishermen in Argentina; he is also such good company that I was half sorry the fish kept us as busy as they did that day, though we made up for it a little on other days.

It was a beautiful day, quite hot, with a brilliant sun and a powerful downstream wind. We drove as far as we could by jeep, then started across the sandy floor of the valley, sidestepping among bunch grass and thornbushes. In half an hour or so we came to the river at the tail of a fine, long pool, just above a sharp turn and a bad rapid, strewn with big rocks and split by an island.

We waded across the tail of the pool without quite the difficulty Tito had anticipated — he had stripped to his shirt and tucked even that up around his shoulders — walked to the head of the pool and then put up our rods. Tito told me to fish the pool while he went on to another a little way upstream. My friend Lee Richardson, who was taking movies, stayed with me, because it was the Long Pool that was expected to produce.

It was an exciting proposition to start into: a new country, new fish of good size and formidable reputation, and a perfect stretch of water. I went at it with the scatter-brained excitement of a novice and a feeling of breathlessness I had thought lost forever in the waters explored in my youth. The pool was at least six hundred feet long, with a shallow rapid breaking into the head of it and a stiff current running through its whole length, especially under the cut bank on the far side. One or two big rocks, also over towards the far side, broke it nicely near the head. There was a comfortable, sloping gravel bar all along my side and, apart from the powerful downstream wind, not a difficulty anywhere.

Foolishly, I had forgotten to ask Tito what fly to put up, but choice didn't seem very difficult. Being a spawning run of salmon, the fish presumably were not feeding much, if at all. But undoubtedly they had spent most of their lives feeding on crayfish, and I had done wonderfully well in Chile on crayfish-hunting rainbows and browns with a large squirrel-tail pattern. I put one on a 9/5 leader and started down the pool with it, throwing well across to the far bank. I didn't expect immediate results — salmon, after all, are salmon — but I was determined to find a fish in the pool.

The first fish came up lazily in the bright sunlight and the clear water, just opposite the second big rock and right under the far bank. I thought I was easing the hook into him perfectly, but there was no resistance on the line and he sank slowly back out of sight. Lee had seen him too, so I had some sympathy.

"What'd I do wrong?" I asked.

"Gee, I don't know," Lee said. "Looked okay to me."

Hooked or not, I felt quite well satisfied to have raised a fish; after all, Tito had said there might be nothing in the pool, and these were salmon. I worked on with the big squirreltail, forcing it across the stiff wind, setting it close under the far bank and delighting in the way it drew down and swept into the current. I don't just remember all that happened or what strange little quirks of fancy made me stay with my bushy brown fly. It raised fish after fish for me, and always I managed to miss them. I told myself I wasn't striking right, I tried waiting, tried hurrying, and always I was using up fish, because none would come a second time. As I worked towards the tail of the pool I began to feel a little anxious. The fish had all come at me close under the far bank, on the swing of the fly, and I wondered if I was letting it come out too fast, in spite of Tito's assurance that they preferred a fast fly.

Then a fish came in midstream, a big fish, at least fifteen pounds, I felt sure. I felt his pull, saw the great silver bar of him turn in the sunlight and sink down. But I still was not fast in him.

"Guess I ought to change the fly," I told Lee.

"Why don't you?" he asked.

"Well, there isn't much left of the pool and at least they're rising to this one. I can always come down again with something else."

A few casts later I had a solid, heavy pull and knew that the fish had taken me properly. Then he was gone, and so was my fly. It was obvious that forcing a long line out across that heavy wind had weakened the leader just above the fly; obvious, too, that it wasn't my morning. I had ruined as fine a pool with as many willing fish as I ever hope to see, and had not a thing to show for it.

We met Tito for lunch and he had two fish of about six pounds each, caught on a Black Ghost. "Little fish," he said. "Very little." But I was glad to see them. They were beautiful fish, bright silver, quite slender, with heavy black spots on back and gill-covers and fins, that seemed rather large in proportion to their bodies. Both were females, with ovaries not more than one-third developed, though I am told the peak spawning month for sebagos in Argentina is March; my own guess from all the fish I saw there would be more nearly mid-April. Both fish had completely empty stomachs.

Tito was very matter of fact about my shameful morning. "After lunch we'll go up where I was. It is a very small pool, but I think there may be another fish there. The fish in the Long Pool won't be hurt — they'll come again. It's a big day — you'll see."

Tito's small pool was at a bend, under a high, steep bank. It was very deep, a confusion of swirling, blue-green water with a wide eddy on our side. I had changed my squirreltail for a slender Iris streamer pattern on a long hook. It caught me a fish almost immediately in the riffle at the head of the pool — a pretty little sebago of two or three pounds which Lee photographed and I released.

Casting across the broad eddy, into the body of the pool, I found I could see my fly perfectly in the bright sunlight. "Work it fast," Tito said at my elbow, and at that moment a big fish came up out of the depths. He missed the fly somehow and it swept past. He turned excitedly to follow it, did not see it and swung back to face upstream again, still just under the surface. I recovered with a gentle roll that set the fly six or eight feet upstream of him and a little beyond him. The moment it touched the water — perhaps while it was still in the air — he started swimming for it. This time his interception was perfect and I let the pull of the line draw the hook firmly into the corner of his jaw. He jumped at once and ran nicely across the pool. Nothing went wrong that time and five minutes later I beached a bright and shapely eight-pounder.

We went back to the Long Pool and I made Tito go down ahead, so that I could watch him. The wind was still very rough, but he was casting beautifully with a French parabolic rod and a forward taper line. He fished his fly with great pulls of hand and rod top working together. "Makes them very excited," he said. "Gets them all upset." A good fish rushed at his fly and missed it. A few feet farther down a big fish took firmly. I started in behind and was into a fish as Tito beached his — a twelve-pound male with a hooked jaw. Mine was a ten-pound female.

We worked on down the pool, hooking fish all the way; my slender streamer hooked every fish that rose to it, in impressive contrast to the bushy squirreltail. I enjoyed the fish because they were beautiful, and because they rose so magnificently, often breaking water at the fly. But I was a little disappointed that they did not run farther and faster or jump very freely.

The first run usually took a few feet of backing beyond the fly line and there would be a jump somewhere in it. Then the fish would come slowly back, almost to my feet, and I would wonder about tailing him or beaching him. A shallow water jump, often right under the rod top, and another strong run just into the backing would be the answer to that. And so it would go, perhaps three or four strong runs and several jumps, then shorter runs and splashing, dangerous struggles in the shallows before the final yielding. I began to feel a little ashamed of having changed from 9/5 up to a 4/5 leader after breaking the fly off in the morning. But it was clear that the fish didn't mind the change and those sudden, close-in jumps were a warning against using lighter gear than one had to.

A big fish had broken water several times close against a pile of brush on the far bank just below me. I called to Lee to have his camera ready, and put the fly across. It was taken with a tremendous splash and I heard Lee say behind me: "Boy, I got it. Now make him jump so we can see what size he is."

But the fish would not jump. He splashed again and rolled on top, and stayed across the river from me, slowly taking out line. Tito was fishing seventy or eighty yards below me and I thought it would be easy to hold my fish up and avoid disturbing him. But the fish still took line, slowly, against heavy pressure. Even when I eased up he would not swim back. It was obvious that I had better work on down, below him. Then Tito hooked a fish and I decided to try again to hold mine.

He still didn't jump or really run. There was not an occasional struggling splash over by the far bank and that steady, powerful pull that made me give line at something a little more than the pace of the current. In the end I had to go on down.

Tito's fish was close to him, so he dropped his rod point and I stepped over his line easily enough. "Hurry," Tito said. "Don't let him too far down the pool. Once he starts into the rapids you're finished."

I think that was the first time it had occurred to me I might lose the fish, and even then I didn't take the idea very seriously. I slacked line so that the fish would hold, went well below him, then waded into the river and put strain on him from downstream, hoping he would surface against it as steelhead and Atlantic salmon have done for me time and again. Instead he came slowly back down to me. I decided he must be tiring and I might as well have a look at him, so I kept him coming. Ten or fifteen feet directly upstream of me he held again. Part of my leader was out of the water, but the light was bad and I could see nothing of the fish. I lifted hard, trying to make him show, but he would not. I slacked away and he still stayed exactly where he was. I stood for a moment, wondering what to do. Obviously I had the tail of the pool blocked off. The only thing he could do was swim back up against the pull of the line and surely that was what he would want to do, go up and find his resting place again. The only question was how to start him. Wade up maybe, and kick him on his way.

I think he moved first. A sudden sharp turn, and a streak past me, down through the shallow water at the tail of the pool and on into the rapid. I blundered after, over the slippery rocks, losing line fast. I managed to steer him on my side of the island in the rapids, but he had seventy-five yards of backing out and was taking more when the line wedged among the big rocks. The 4/5 gut held for a moment, then broke.

I told Tito: "He was either a very big fish or he was foul-hooked."

"He was both," Tito said. "You shouldn't have tried to hold him."

I caught the Argentine sebagos in other places and on other days, but I never again hit a day like that one; a day when, in spite of every mistake I could make, fish insisted on being caught. I have regretted ever since that I did not try a dry fly in the Long Pool, a good bushy hair fly, perhaps a Wulff or one of the McKenzie patterns. I am sure the fish would have taken it readily. But the idea didn't enter my mind, perhaps because I was too busy, perhaps because the powerful wind gave me all I wanted to do with the wet fly.

But even without testing the dry fly, it was a day of many lessons. Perhaps I need not have learned them all the hard way. It was reasonable to be surprised that non-feeding fish should take as freely as the hungriest of feeding trout. But most people know enough to change a fly before six or eight good fish have come short to it. And it takes a really dull mind not to realize in advance of the event that fish migrating downstream will have no inhibitions about breaking out of the tail of a pool down to the next one. Fish get fooled by habit: fishermen are supposed to be smarter than that.

15
Salmon of the Vatnsdalsa
(1969)

A wide, flat, green valley with red-roofed farm buildings and a bright, clean river; steep green slopes on either side climbing quickly to black rock outcrop and snow-filled gullies; low clouds along the mountaintops, shifting valley winds, break of blue sky and sunlight; handsome, pale-gold people of great antiquity; arctic terns, golden plover, and black-backed gulls; wild flowers blooming in profuse succession. All these things make up northern Iceland in mid-July — all these and the salmon.

When my friend Bill Gregory called from Minneapolis in May, I had only an uncertain impression of all this, but I thought immediately: Atlantic salmon, small but lots of them, attractive streams, country something like the Canadian Arctic. "It's a small stream," Bill said. "Only twenty miles long, limited to four rods. Fly only and all wading, no boats. I have it for mid-July and we're a rod short. Can you come?"

I knew I was going, of course, but even in this age of easy travel, a journey across a continent and half an ocean takes a moment or two of thought. Then, too, I am magistrate and family court judge for some fifteen or twenty thousand people in a ten-thousand-square-mile area — very law-abiding people, naturally, but even with an experienced deputy and a capable court clerk I find myself sometimes setting trials as much as two months ahead. So it was necessary to look over the court calendar. I did so, and booked an airline passage to Iceland for July 10.

I found myself in Iceland only one day later than I had expected, checking into the Saga Hotel in Reykjavik in broad daylight at 2:00 A.M. The lovely city was silent, not a car or a person moving in its streets, the elegant buildings unlighted, even the flags quiet on their masts. I drew the heavy curtains firmly against the daylight, and then slept gratefully in the silence.

The river we were to fish is the Vatnsdalsa, one of several flowing into Hunafloi, the great fjord that cuts widely back into the northwest coast of Iceland. It is a six-hour drive, north and east of Reykjavik, by a gravel road that skirts the edges of two west coast fjords, climbs over a low divide, and then drops down across three or four valleys and their rivers to come to Flodvangur, the comfortable lodge in the Watnsdalur, the valley of the Vatnsdalsa. It was a fine introduction to the country. We passed green farms in the valleys with good herds of dairy cattle; on the unfenced tundra were the beautifully formed small native horses, and native sheep with long, silky fleeces that rippled in the wind. Wild swans were paired, nesting in the potholes; eider ducks, terns, and black-backed gulls rested along the shorelines; and in places the polar ice was piled in formidable blue and white drifts for miles.

Iceland has about sixty salmon streams, which are managed almost entirely for rod fishing. Small set nets are permitted for limited daily periods on a few large glacial rivers, where the silty water makes rod fishing unproductive, but there is no saltwater net fishing. Rod fishing rights are owned by the riparian farmers, who usually work together to lease a whole river or sections of a river to a club or an individual, who may then sublease by the day, month, or season. The worth of a river is determined largely by the number of rods it will support and this is decided by the Directorate of Freshwater Fisheries on the basis of past catches. The salmon season on most of the rivers in Iceland runs from June 15 to September 15, and fishing is permitted for only twelve hours each day.

The Vatnsdalsa, as Bill Gregory had told me, is a four-rod river — that is, only four rods may fish for salmon at any one time, between the hours of 9:30 A.M. and 9:30 P.M., and with fly only. There is no limit on the catch and the fisherman is expected to kill all the fish he catches, which are then frozen and shipped to commercial markets. In this way the river's record is accurately known from season to season and the number of rods can be controlled accordingly.

We had come to Flodvangur without seeing the river itself. The lodge overlooks a small shallow lake (Flod) through which the river flows. Below the lake are two highly productive pools, Landslide and the Peat Pool, just above tidewater. Immediately above the lake the stream is slow and flat, meandering through peat-bog meadows and around bright green islands for three or four miles to the tiny church that serves the valley. Salmon do not hold anywhere in this stretch, which is known as the Trout Water and is full of Icelandic char and brown trout. A fisherman on the Trout Water is not reckoned a salmon rod and is encouraged to take as many trout and char as he can catch, in the hope of reducing predation on the young salmon.

Above the Church Pool is another twelve to fourteen miles of river, divided into three beats, each with several fine pools, each with its own special character.

All this was quite vague in my mind on that first evening, compounded of talk and maps and a list of the names of the pools — Dalfoss, Green Banks, Grettis, Char, Grimstunga, Junction, Red Braes, President's, and Corner Pool were a few of them. The fish were late, we were told, nearly two weeks late because of the cold, late spring and the ice floes we had seen piled along the beaches. They were beginning to show quite well in the Peat Pool and Landslide, below the lake, but very few had yet passed through to the upper beats. In front of the lodge the little lake was flat calm under the grey daylight of midnight; low clouds hung along the dark faces of the mountains across the valley, golden plover called from the rough land nearby. It was enough to go to sleep on a new country, a new river, new birds, unfamiliar fish, and a thousand questions that the next week or ten days might answer.

I sometimes think the modern fly fisherman has far too wide a choice of gear. Our host, Major Ashley Cooper, had recommended twelve- to fourteen-foot double-handed rods, double-hooked flies sizes 6 to 9, with a few larger ones up to 2/0, and leaders of nine- to sixteen-pound breaking strain. I have used the big rods enough to know that I prefer single-handed rods of nine feet or less if they will do the work at all, so I had brought along two long rods and three single-handers. I had an ample selection of flies of all types and sizes, together with an undesirable curiosity about the effectiveness of steelhead patterns for Atlantic salmon, and a slightly more sensible curiosity about the responsiveness of Icelandic salmon to floating flies. North American salmon like them, European salmon apparently do not; would Iceland fit between? Finally, I had sinking lines, intermediate lines, floating lines, and one "wet-head" line. This was perhaps a pardonable confusion, since the use of double-hook flies suggested to me that the fish might be taking deep, while the small sizes suggested they might be very close to the surface. I would have been better off to leave most of the stuff at home, but a fisherman always has his doubts and theories, and once in a while something unusual does pay off.

The river was far more beautiful than I had expected. Between the pools one could wade it easily almost anywhere, yet the pools themselves were large enough to call for a good, long cast and plenty of searching. The valley is practically treeless — a few dwarf birches fenced off from the sheep, here and there sad little plantations of ailing pines, less than ten feet high, a few thrifty rowans in the churchyard — but it was green everywhere with a richness of short grasses and sedges and mosses; in amongst the green and even on the gravel bars, was a profusion of wild flowers, little arctic creatures making the most of the short northern summer. On both sides of the river were handsome white farm buildings with red roofs — the oldest farm in the valley, we were told, had been there for over a thousand years. The valley's first quarrel, one which ended in murder and which is recorded in the sagas, was over salmon rights.

Above the Trout Water the valley continues flat and wide and green, but the river is faster, with alternating pools and rapids over a gravel bed that clearly shifts and changes in the winter floods. Just below the Grimstunga Bridge is the Junction Pool, where the little Alka River comes in from the west. Junction is considered one of the best pools on the river and had produced two or three fish the week before we arrived. It is a simple, straightforward pool, with a fast run at the head, good depth, a nice spread toward the tail. It was the first pool we fished and someone fished it at least once every day we were there, but it did not yield another fish until the very last day.

Above Grimstunga Bridge the valley narrows and the hills crowd steadily closer until the river is flowing through a gentle canyon, still with steep green banks between the rock outcrops. At the head of the canyon is a fine fall, about twelve feet high, and below it a boil of turbulent water between high rock walls that spreads into a deep pool with a wide tail and a good shallow glide under the far bank. I saw one salmon jump at these falls, but it was a frail attempt that did not carry him more than half their height. It is possible, though, that salmon do pass them under high water conditions, in which case they would reach another three-quarter mile of fast, rocky water before coming to a much higher fall.

This pool is known as Stekkjar Foss, or Dalfoss, and always holds fish, some of which drop back to lie in the glide at the tail. The first time I saw the pool Jesse Oppenheimer of Texas was fishing it and had been for an hour or more. I saw him first from a distance, but even so it was evident that there were fish in the tail of the pool and he was approaching the limits of frustration. His fly box was open on a rock on the near side of the pool. After every half-dozen casts he would return to it and change his pattern. By this time, I had come up to the pool and could see four good fish lying almost abreast between two big rocks on the far side, in three or four feet of water. Jesse came to his final change and put up a dry Grey Wulff as a gesture of desperation. His first cast was short. I forget about the second. The third was just right and a fish took it perfectly. It ran around the pool for a few minutes, then went out through two hundred yards of white water chutes to where we landed it in the next pool down.

I felt at that point that life was going to prove pretty simple. A nice hair-wing floater would take fish without any trouble. A day or two later I watched another fisherman attempt the same thing, with the fish in much the same position. They showed interest all right. But it took well over a hundred casts and several changes of fly to bring one up — this time to a fly with bushy white wings and a black body, known as a Pass Lake. When the pool had settled down and one or two fish had moved back into position, I tried it. There was little interest until I put on a wet Blue Charm that dragged over them at enormous speed. A fish chased the second cast ten or fifteen feet downstream with half his side out of water. He didn't catch it, so I threw back. He chased again, in exactly the same spectacular way, but this time he had it securely and in due course was landed.

There are several ways of taking fish in the Dalfoss Pool, not the least of them by sitting or lying on a bluff some thirty feet above the water and fishing a downstream wet fly. It is always fun and especially tempting when fish are scarce in the upper reaches of the river. But I suspect the time would usually be better spent in searching through the other ten or twelve fine pools that lie between there and the Grimstunga Bridge on Beat Number One.

Fishing the upper beats in this way was a delight. I was only some ten days out of the hospital and not feeling too strong, so I did more watching than fishing. But there was always a fish or two to be found somewhere. In Green Banks, Long Pool, Red Braes, or one of the others, and we were never blanked. The weather was uncertain but generally warm and often sunny, nearly always with a changing wind. In the canyon area it could change through all four points of the compass in the duration of a single cast, calling for some pretty sharp adjustments if things were to go smoothly. There were always new flowers to examine, sometimes hanging banks that were a complete garden of grasses, sedges, mosses, and star-like flowers, sometimes violas and harebells and arctic fireweed scattered in the grey gravel. The birds were lively and beautiful: arctic terns everywhere performing wonders of flight, the magnificent whooper swans sweeping up and down the valley; whimbrels, practically the same bird as the Hudsonian curlew; mallards and goldeneyes and harlequins; snow buntings, redshanks, cock plover, and abundant golden plover.

I could truthfully say that I needed no more to keep me happy. But it was a slow way to learn about the fishing. When fish are scarce, one experiments endlessly and proves very little. It was in the Peat Pool and Landslide that we learned, when the rotation of beats took us down that way. Fresh fish were moving into the Peat Pool from the sea every day. One would see them passing the long, shallow rapids between there and Landslide and even passing above Landslide to disappear into the little lake. It seemed impossible that they were not passing on out at the other end, through the Trout Water and into the upper pools. But they were not.

Peat is a very large pool, two pools really. The shallow rapid comes in at the head on a bend. Half the flow runs in over a shallow riffle on the west side, while the other half turns along the steeply cut east bank, and the two come together a long cast down from the head of the pool. The flow remains strong and quite wide along the cut bank for another hundred yards or so and it is all good water. A large eddy, fairly deep but wadeable in most parts, forms on the west side. The second part of the pool starts above where the current spreads to reach the west side again and continues from there for a considerable distance. How far, I am not sure; I raised fish as far down as I went, but it is possible some of them may have been moving through rather than resting.

The first day I fished the Peat Pool and Landslide it was bitterly cold — the only really cold day of the trip — with a wind blowing twenty-five or thirty miles an hour straight down from the polar ice. I was using an intermediate line, ungreased and therefore sinking, and rather large flies. I think I lost all of the few fish I hooked and I came away with no clear ideas of how the pool should be worked.

Three days later I fished it again for an hour or so on a pleasant morning. By this time I was using a floating line with a No. 6 fly and had pretty well settled to the idea that the only pattern needed was a Blue Charm. Thunder and Lightning, Black Doctor, Lady Caroline — any small dark fly would do equally well; what mattered was the size and how the fly was worked. I was casting well across and about forty-five degrees downstream, letting the fly come through the current without artificial movement and picking it up quite fast as it came into the slacker water. I hooked a fish almost as soon as I started and lost him at the net ten minutes later. I went back to the head of the pool, hooked another fish almost at once and killed him.

Landing fish in the Peat Pool was almost a ritual. One left a good large net about halfway down the pool, at the edge of the eddy. On hooking a fish in the upper part of the pool one moved it immediately out of the water and downstream, to lead him away from the good water. Most of the fish were a little slow in responding to the hook; they would hold rather quietly for a moment or two, long enough to allow the fisherman to get out on dry land and start down. Then, when the pressure came on in earnest, they would make two or three strong runs, well down in the backing, sometimes jumping clear of the water, more often surging out in powerful lunges that plowed the surface white and slowed the run not at all. One fourteen-pounder ran me five times, each time at least forty yards into the backing.

After this they usually came in close and one felt one had them — at least I started out thinking that way. One could lift them, roll them, even check an incipient run usually within the length of the fly line. But leading them into the net or even onto the beach was another matter altogether. There was always a reserve of strength and on a short line it was always dangerous. Netting fish like these on one's own was an intensely exciting affair. The temptation was to attempt it too soon instead of waiting out the slow, powerful resistance that could so easily turn into another run. Any attempt to net from the tail, even if the fish was on his side, was certain to produce another run into the backing and probably out of the pool. Leading a fish over the net and hoping to lift as one would for a trout was equally futile; even if his head was over the centre of the net at the moment of the attempt he would sense the movement and be gone. The one and only way was to lead him head first into the net, when his violent reaction would drive him straight into it. Even this was not without its hazards. One fish I led in this way hit the net so hard he took ring and net right off the handle. It was over deep water. I made a grab for the disappearing net and fish and slipped onto the seat of my waders. A second grab, still from a sitting position, was successful, though I got pretty wet.

For me these final strenuous moments added a new dimension to the fishing. I am used to fish that tire themselves on a relatively long line and can usually be controlled when they come in close. In Iceland I reminded myself with every fish I hooked that those last, long moments were ahead.

It would be easy to write much more of the Peat Pool, which is both an easy pool and a great pool. I never saw it without fish and never left it without knowing I could have taken another fish there. And I believe I solved it in the end. One morning I lost four fish in succession, each one right at the net, on a No. 6 Blue Charm. Then I had the wits to change to a No. 8 and killed the next three fish I raised.

By this time I had settled down to gear that I considered ideal for the job: a nine-foot, five-ounce Winston glass rod, with no metal ferrules, No. 10 weight forward Air-cell Supreme, Hardy St. Aidan reel, leaders tapered to 9/5, and a No. 8 Blue Charm or Silver Blue. This, of course, is greased-line equipment. But the fish seemed to want a fast fly. I made my cast to the far bank, quartering downstream, let the fly settle briefly, and brought it across the current with a series of eighteen or twenty short twitches, taking in perhaps three or four inches with each one. If the fish were really on, this is where the take would come. As soon as the fly came into slower water below me I recovered my seven or eight coils of running line and cast again. If the fish were going off the take and following round, the faster recovery would often persuade them to take hold.

This method of fishing meant that the fly was always practically in the surface film — at times I thought too much so, because I could see its wake, and in Landslide Pool, which is very slow and smooth, I went down to 1x leaders to avoid this.

The last time I fished the Peat Pool was a beautiful Icelandic morning, with little wind and a lightly clouded sky. There had been some rain and the river had risen two or three inches during the night. The Major dropped me off there on his way up to Landslide. "It looks like a good day," he said. "Don't work too hard. Sit down and rest for a while when you get a fish." I made up my mind to do exactly that.

It was the first time I had had the pool entirely to myself and I did everything deliberately: parked my net at the edge of the eddy, checked my gear carefully, walked slowly up to the pool, talking to the nervous, scolding terns to try and persuade them I had no designs on their downy young. Up at the head I sat on a rock and watched the pool until I had seen two or three fish roll, then slowly crossed the riffle and put out line until I was reaching the cut bank. I was into a fish almost at once and landed him safely — a handsome cock fish of fourteen pounds. I admired him, photographed him, and looked at my watch. It was 9:50 A.M.

I walked slowly back to the head of the pool, found a comfortable rock, and sat down with my back against it. For twenty minutes I watched the mountains, watched the terns as they chased off some ravens, watched the slow flight of seven swans low across the tundra on a line where I had never seen them before. Then I went back into the pool and in fifteen minutes landed another fish.

By 12:30 I still had not passed the upper part of the pool and I had five fish on the bank, all between twelve and sixteen pounds, without losing a single one. I wanted no more from the pool and sat down to wait for the Major to come down from Landslide.

There was time to reflect on the calmness and precision of the whole morning. There had been no fumbling, no wasted effort, and the tiny fly had held firmly even through those final dangerous struggles. Most of it was due to the fish themselves. Every rise had come on the swing, most of them toward mid-current. They were quiet, deliberate rises, plainly visible in a shouldering movement that seemed scarcely to break the surface. I did not strike to the rises, but waited to feel the fish, then set the hook as far as possible from the side.

Landslide in the afternoon adjusted my perspective a little. It is about as different from the Peat Pool as it is possible to be. It lies just below the lake, close under one of a number of little round hills, about fifty feet high, left by volcanic or glacial action. It is a small pool, perhaps fifty feet wide and two hundred feet long, no more than six feet deep and with a very slow, smooth current. Fish lie well in it and one can see them plainly from the shoulder of the little hill or even, if there is no wind, from the level of the pool itself. With a strong ripple on the water it is easy to take fish there, but in a flat calm it can be quite difficult.

It was flat calm when I arrived there that afternoon, but I decided to make no change in the method of the morning, except for keeping low and keeping my distance. Two fish followed the fly and surged at it near the head of the pool, but I felt neither. Then near the tail I hooked a fish that went completely wild. He jumped and ran and jumped and ran again through every foot of the pool, no matter how I tried to calm him. Fish boiled and twisted away from him in every direction and one even jumped clear out of the water. He wouldn't leave the pool and I couldn't lead him up away from it toward the lake. When I got him under control I landed him just below the tail.

I rested the pool, tried again, and lost two fish in quick succession. Then I tried a floating fly, working up from the tail. The fish turned to it, made uneasy movements under it, rolled out after it had passed them. One fish came up beautifully from a distance, drifted back just under it for three or four feet, then turned away. In the end I had to admit they were not going to take it, though I am sure that with slightly different conditions it could be very effective.

Landslide Pool makes up less than a tenth of the width of the river at this point. The rest is shallow rapids except for a narrow run some four or five feet deep toward the far side, called Bjarnastadir Hylur, where fish hold fairly well. I had searched for it once or twice, but was never too sure whether I was fishing it or wading through the middle of it, so I decided against going out to look for it again. Below Landslide is the long, shallow rapids that goes all the way down to the Peat Pool, half a mile away. There are small potholes in the rapids where fish hold briefly and I had seen a fish roll in one of these, by a large tuft of grass about seventy yards below Landslide.

I decided to try for him and was wading out to get in position when I saw a good fish lying in less than a foot of water about fifteen feet away. I froze, then cautiously withdrew, moved upstream, and cast the little Blue Charm over him. To my surprise he took it at once, and was firmly hooked.

What happened from then on was totally without precision, deliberation, or any other quality except excitement. The first place he went was into the hole by the tuft of grass, exactly where I didn't want him. Even so, I tried to keep him there and tire him a little, but he went off across the rapids and into a maze of half-submerged rocks hung with trailing green weed. I scrambled back to shore, picked up the net, and went after him. He eased up a little and I steered him back among the rocks somehow. He was far from ready for the net, but I began to wonder just where I could hope to net him. The best chance seemed to be just below Landslide or up in the pool itself, if possible. But my fish thought otherwise. He came back all right, to within thirty or forty feet of the rod tip and swirled sulkily as I tried to lift him. Then he went off again.

He was past the grass tuft and forty yards into the backing before I decided to follow. Even then he was still running and the glass rod was singing to the strain — a disconcerting metallic sound I was still not entirely used to. Not far below us was a wide, shallow ford and I didn't like the idea of his flopping around on that at the end of a hundred yards of line. He turned short of it and I managed to bring him back; he was obviously tiring. Again I looked for a place deep enough to get the net under him, but there was nothing. I dropped the net and decided to beach him.

He was not ready for that. When I tried to lead him in he

went off again, half-swimming, half-flipping and sliding over the shallows. Soon he was onto the ford and I had to ease up and let him find his own way over it. There was a little more water below, but nowhere a shelving place where he could be led unsuspecting to his fate. I tried it four or five times only to have him flip and scramble away. When I finally managed to keep him quiet for a moment it was a good four hundred yards downstream from where we had started, and even then I had to tail him and swing him up onto the grassy bank.

The next day was my last day of salmon fishing. Only three of us were on the river — the Major, Bill Negley of Texas, and I. The Major suggested I try the Peat Pool again, but I wanted to see the upper pools once more; after all, they were the real river, with the spirit and weather moods, the beauty and remoteness of Iceland about them. I knew that Bill Negley, who is an excellent fisherman, had been up there the day before and found only one fish. But there had been a rise of three inches or more in the river, and the fish had to go through some time.

So Bjarni and I started out in the Land Rover. Bjarni is a young school teacher from the great peninsula that swells northward on the west side of the Hunafloi. He knows all the pools and the easiest approaches to them; he is calm and good with the net, and very good, quiet company. "Where are we going?" he asked. "To Dalfoss?"

"No," I said. "We would waste too much time there. We'll try the Junction, then go straight on to Krubba and Green Banks and fish all the way down."

I started at the head of the Junction, resolved to fish steadily and fast. A fish rolled near the tail soon after I started. When I got down to him he took the fly on the swing and was duly netted by Bjarni. There were no sea lice on the salmon, but he was very clean and bright, and I was sure he had not been up long. We left the pool without another cast and started upriver.

Unwisely we stopped on the bridge over the Alka River and looked down into the pool below the canyon. For the first time in all the times we had looked there was a fish resting there, and a good one.

The fish was lying in quite shallow water near the tail of the pool; there was only a slight current and the surface was perfectly smooth. I crept down under the bridge and threw the little Blue Charm to him. Bjarni told me he moved to it. He moved to the next cast also, then no more. I eased a few feet downstream to change the angle and again he moved twice and no more. I changed to a larger fly. He paid no attention to it until I dropped it behind him, then he turned after it and turned away. I reeled in, crept down past the fish, keeping under the shadow of the high bank, and asked Bjarni for my other rod, the little Pezon-Michel that was ready with a floater.

The fish flickered his fins and moved slightly at the first drift. At the second he made an abrupt, agitated circle, but did not come up. He managed to appear steadily less interested through the next half-dozen casts. I thought of changing the fly, but instead I told Bjarni: "We're wasting as much time here as we would at Dalfoss. Let's go." Bjarni grinned and we went.

Below Dalfoss is the Trout Pool, which does not hold fish very well, then the Non Pool and the Dog Pool, both of which are small. Below these are Krubba and Green Banks, two of the finest pools on the river. Krubba lies under a high rock face on the east bank. The current enters rather narrowly at the head, turns sharply against the rock, slides along it, and spreads to a broad, deep pool. The west bank is smooth and grassy, but climbs steeply.

Green Banks starts about a hundred yards below Krubba, with a formidable run-in of white water that swirls and settles among several big boulders. The rock face is on the west side and the current flattens and eases all along it as far down as a big, round, reddish boulder in midstream, which is about the end of the holding water. The west bank slopes very gradually from the water's edge, a rich green garden of grasses and wild flowers with a steeper slope, still green, rising behind it. A place to dream of, to lie in the sun and listen, to make love; far too perfect, certainly, for the needs of a simple fisherman.

I fished Green Banks first and saw a fish roll by the rock at the tail almost as I started. A fish touched my fly near a corner of the rock face, but would not come again. The fish at the tail rolled at my fly but did not touch it. So I went up to Krubba, but I knew I was not finished with Green Banks.

In spite of its looks and reputation, Krubba had not yet shown me any sign of a fish. I had fished it several times, and watched others fish it, with floating and sinking lines, large and small flies. I had looked into it from the top of the rock bluff and seen nothing. I fished again now, with deep faith, carefully and with my best technique. Nothing moved. Bjarni had gone upstream to look over the Dog Pool, but there was nothing there either, so we had lunch to give Green Banks a longer rest.

It was a pleasant day of clouds and occasional sun and twisting, uncertain winds. I very much wanted a fish from Green Banks and hoped to find one among the big boulders at the head. But the first fish took at the start of the rock face and right against it. He moved quietly into midstream and let go of the fly. Half a dozen casts later a second fish did exactly the same thing and I began to wonder if the little Blue Charm was right after all. But I fished on down to the big boulder at the tail, raised my fish there, and hooked him solidly. Bjarni landed him for me and I knew, sadly, that it was time to go on.

The Grettis Pool, Bjarni told me, is so called because it is the place where the Viking killed the ghost. Bjarni reads the sagas as he waits beside the river and says they are very gloomy and tragic; and in that northern valley it was easy to imagine the sources of old legends and feel them all about you. The valley's first settlers, in that farm of a thousand years ago under the frowning mountain, would have felt the challenge of the unknown hinterland. Sea people, huddled through the long, dark winters in the strange land, they would have told stories. And in the long daylight of summer some brawny younger son would have shouldered his spear and gone forth to meet the challenge, bold yet fearful. Certainly he might have met, among other terrors, a ghost by the Grettis Pool.

For the Grettis Pool lies just where the valley begins to open up to farmland, though still a dozen miles above the site of that first farm. A foaming rapids dances briskly into it, strikes hard against a low rock outcrop in the west bank, swirls from there to spread into shelving tail and back eddy. A tiny pool, but deep and good to hold in. The sun was bright as I came to it — too bright, I thought, as a fish rolled lazily at my fly and missed at the edge of the current just below the outcrop. He missed on the next swing, too, but had the third one firmly. Bjarni landed him fifty yards below the pool and I went back to finish out. There was another fish at the tail, but he missed the fly and would not come again, so I left him.

I fished on in bright sunlight through Slope and the Long Pool and nothing moved. It was still sunny when I came to the Char Pool, where Bjarni said yesterday a good fish had rolled near the big rock above the tail. Char is a meadowland pool, swift at the head but with a brown earth cut bank on the east side. I fished it carefully because it holds well. A cloud came over the sun just before I came to the rock. The fly swung over and past it, then came to a solid stop. I hesitated a moment, then stirred things up. The fish went away fast and, to my surprise, he was small, probably a grilse. I kept him in the pool and finally brought him up to where I could see him in the water. The orange belly showed up plainly — a char of about five pounds. I treated him roughly from then on, but even so he ran twice into the backing before I could net him. Icelandic char are surprisingly strong fish and the small ones jump freely.

I was not satisfied that the rock had been properly covered and my next cast hooked a good fish just on the far side of it. He fought hard and dangerously as I tried to work him up the pool, running back toward the shelter of the rock every time I gained a little on him. In the end he ran out of the pool and into the shallow rapids below. I chased after him and about two hundred yards downstream we found a deep enough place for Bjarni to slip the net under him. He was well over fifteen pounds.

By this time it was getting late and I had not covered nearly as many pools as I had hoped to, so we ran on down toward the other beats. I fished the Junction again and rolled a good fish near the tail, raised and felt a fish in Red Braes, fished Pollarnir without moving anything, and came to President's Pool, under the dark clay bluff, at 9:15 P.M. It was too late to fish the pool properly, but I swung a fly quickly down it and saw one fish clearly against the white clay bottom on the far side. Bjarni waded out to stand beside me, then we saw two more fish in the shallow water right at our feet. Obviously they were moving up.

We had fished until the stroke of 9:30, but were still first back at the lodge. The Major and Bill Negley arrived shortly after with seventeen fish from the Peat Pool and Landslide. Bill had hooked the last fish of the day, a magnificent twenty-one-and-a-half-pounder which had taken over an hour to land.

When I am asked what I think of the fishing in Iceland, the only sensible answer I can think of is: "I wouldn't want it any better." True, we may have been a little unlucky in the late arrival of the fish. It would have been wonderful to have had fair numbers in the upper pools all through, instead of just on the last day. What else should one want? Larger fish? Perhaps, but I'm not at all sure. Larger fish would mean fewer fish and probably less willing fish. However I assess them, the last three or four days I spent on the Vatnsdalsa were as exciting as any fishing days in my life.

In one way we may have been lucky in the late arrival of the fish. Generally, I understand, grilse make up some 50 per cent of the catch. We took hardly any — perhaps four or five in the whole time. Most of the fish we caught seemed to fall roughly into two size groups, the smaller ones about thirty inches long and weighing nine to ten pounds. The larger ones thirty-three to thirty-five inches and weighing from twelve to sixteen pounds. One or two eighteen-pounders and Bill Negley's twenty-one-pounder may be in still another age group. The record for the river is, I believe, twenty-eight pounds.

So far as I can recall I saw the rise of every single fish I hooked and this to me is the ultimate in quality fishing. Quite a number of fish rose two or three times on the same swing of the fly, to be hooked on the final attempt; this is exciting. A good number rose handsomely and were not hooked; this is frustrating, but also exciting. For the most part, though, they rose deliberately and solidly. That they came best to such a small fly was yet another satisfaction.

Not one fish I hooked was easily landed. Some jumped more than others and surged more violently, some ran farther and more often, but that is always the case. Every single fish had that dangerous reserve of strength at the end. Let him get his head down and he would run again, hard, at first sight of net or first feel of shallows. To land any one of them without taking half the day to do it meant putting on all the strain the tackle would stand — and being ready to let go instantly if the strain became too much.

Iceland is a beautiful country and a romantic one. The people, if one must generalize, are friendly, charming, and quite remarkably handsome. The summer climate is as pleasant as a fisherman could want, except when the polar wind blows, and there may even be a little more sun than is good for the fishing. There are no mosquitoes or biting flies, only a blundering creature that looks and acts like a blackfly except for the bite. The wild flowers and grasses are beautiful. The streams are varied and challenging, clean, and unlittered.

It would be nice to say that the future is secure. Perhaps it is. The streams are well-managed and the government is fully alert to the value of the runs. More intensive management would be possible, but at present there seems little need for it. The salmon disease that is plaguing British and Irish stocks has not yet reached Iceland and one hopes it never will. But Danish fishing in Greenlandic waters and on the high seas may be a more serious threat. No salmon runs can stand up in the face of uncontrolled fishing, and there is no really accurate or effective means of controlling a high seas salmon fishery. One can only hope that the nations which reap that shortsighted harvest will have some change of heart and conscience. Without seed and seed bed there can be no harvest.

16
On the Trout Water
(1969)

We had only a short day left before the taxi came to take us to Reykjavik. I knew I had fished enough especially in the last three days when the salmon had been moving upstream and taking readily as they settled into the pools. But Bill Negley hadn't been fishing as long as I had. "I'd like to try the Trout Water," he said.

The Trout Water is a five- or six-mile stretch of the Vatnsdalsa, a fine little salmon stream in Iceland. It runs from the shallow lake above the sea pools up to the Church Pool, just below the tiny red-and-white church that serves the valley. It is slow and quiet, winding through the meadowlands and islands of the flat valley floor. It has no well-defined pools, scarcely any change in its smooth way except for a few runs of slightly faster water where a sandbar has built up or a grassy island turns the flow. The salmon do not lie there; they simply pass on through to the sheltering pools above. But there are good brown trout and Icelandic char.

Bill's suggestion suited me well. I planned a pleasant, easy day in the green meadows under the mountains, without waders, perhaps trying to get pictures of the whooper swans that worked among the islands. I took a trout rod along and a box of flies and hoped for a suitable challenge that could be met from the bank. After all, Dick Vaughan had fished a whole day there earlier in the week and surprised himself. He had returned all his fish — to the mild annoyance of our host who believed all trout and char should be slain forthwith to protect the young salmon — but had counted among them brown trout of seven pounds and four pounds as well as many char and lesser trout.

It was a cloudy day with a cold wind and showers that settled to steady rain before noon. The swans were feeding on one of the islands. They let us come down the slope past the farm from the road, across the meadows and almost to the river bank, then took wing still at a great distance and moved farther downstream. I was disappointed, but remembered they had passed close in flight on several other days.

The stream was broad and very flat where we had reached it, but the bottom was uneven clay and a few weeds, suggesting holding spots for feeding fish. Bill, who is a strong and energetic fisherman, went off upstream to search for runs that promised more. I wandered along the bank, looking it over. There were occasional rises, hard to see, not repeated. I put up an orange-bodied Humpie floater and covered one of the rises. The fish took well, a char of fifteen or sixteen inches; even at that size, Icelandic char are brave little fish, and this one ran into the backing and jumped several times before I could bring him in and lift him onto the bank. Bright and clean, silvery gold, with orange fins and belly. A second rise produced a twin.

I passed a little round island with cotton grass in bloom, where a mother redshanks fussed prettily over her almost fledged, water-borne young. She hid them quickly and drew me away. I was thinking of the big brown trout Dick had returned and changed the Humpie for a Silver Brown, lightly dressed on a low-water hook. Still wandering along the bank, thinking of the swans more than anything, I pitched the new fly idly across. It took a small brown trout, then almost immediately another. I noticed it had passed over one or two rises without drawing attention. Bill caught up to me and we compared notes. "I think the Silver Brown may be selective for browns," I said. "It will keep you from tangling with char all the time." He liked the look of the fly and took one.

Our downstream progress was stopped by one of the drainage ditches I had forgotten about when I left my waders behind. Four or five feet deep in the peat with two feet of water over the bottom, it was an effective barrier unless I chose to get wet. I decided to go around, and Bill, not caring much for the look of the stream at that point, came with me. Another family of redshanks fluttered and swam ahead of us, hid somehow, and were behind us. The ditch came to the slope of the hill and turned along under it. There were golden plover all over the drier ground. Along the ditch a goldeneye fussed with her dark little brood. Snipe and curlews on the wet ground, more plover on the earth thrown up from the ditch, a few terns hovered overhead even here. We came to a crossing and back to the river again, well below the swans.

Bill liked the look of it, waded in, and quickly took a nice brown of two or three pounds. I studied the swans, distant white specks with the green lower slopes of the mountains rising impressively behind them. As soon as I turned towards them two long white necks went up.

"They're spooky as hell," I said.

"Try keeping low and walking straight towards them," said Bill, who is a great and wise hunter.

I tried, ran into swampy ground, then wetter than just swampy. I dodged around it, found the same again, then more of the same again. A straight course was out of the question and more of the long, white necks were now stretched up. I glanced back towards Bill, fishing a run along a sandbar, and saw him hook another good trout.

A movement overhead drew my attention. Two large, grey birds chasing a smaller one. Birds are always chasing birds, and even people, on these subarctic nesting grounds, and for a moment I didn't recognize what I was seeing. Then I knew this was deadly pursuit: two hook-billed skuas and a tiny snipe. The wild flight went on against the storm clouds, twisting and violent, incredibly swift, and I dared hope for the snipe. Drop, I thought, suddenly, make your little body like lead, then reverse on them and slope down into the sheltering grass. But the chase turned upwards instead of down. A moment later there were only two birds in the sky, the snipe a bulge in the beak of one of the skuas.

It had all been so swift, so relentlessly dramatic that I had had no time to call to Bill. I called to him now and he turned in time to watch the big bird circle back and land on the sandbar behind him. It stood there, scarcely moving, while a few pale feathers drifted away from the pale-breasted body in its beak. Its mate, following lazily, landed nearby on the sandbar. The first bird moved closer, set its prey on the sand and backed away. The second bird began to feed.

It was raining quite hard now, a cold rain driven by a cold wind out of the north. Magnus sounded the taxi horn from beside the farmhouse up on the hill. It was time to go.

17
Big Floaters for Western Trout
(1952)

Several good fly fishermen, visitors or newcomers to the Pacific Coast, have recently told me of their deep disappointment at the reluctance of our native rainbows and cutthroats to rise to floating flies. This complaint never surprises me. I have fished Pacific Coast waters for nearly twenty-five years and am still learning, every season, some very elementary things about the use of the dry fly for rainbows and cutthroats. From the first year I fished coast waters I have caught fish on floating flies; but for many years I did so with a dissatisfied feeling that I was distorting things a little and that rainbows and cutthroats, at least under conditions commonly met with out here, are primarily fish for the sunk fly.

There is some superficial truth in this, enough to be thoroughly misleading. Dry-fly technique was originally developed and brought to perfection in clear and shallow streams, for the taking of brown trout. The brown trout hovers in wait for his food, usually very close to the surface. Being fairly static in his habits, he clings to a single pool, a preferred lie in that pool, a single holt when resting or frightened. He is a cautious, educated feeder in those smooth streams, and much of his food is small May flies, small sedges, small midges.

As against this the Pacific Coast cutthroats are restless and migratory fish; they prefer strong water of some depth; they rise from a depth or a distance or both and with more attack. They feed less selectively, and generally more predaciously, responding readily to movement; and the insect and other life offered by their native streams is generally larger, though often similar to the feed of European streams. Finally, in spite of the more aggressive and positive nature of the western trout's response to floating food, there is less real confidence and certainty in it than in the quiet and easy rise of the brown trout.

In this catalogue of differences is the answer to many of the disappointments and much of the dissatisfaction. Most of us have come to western fishing with preconceived notions based on knowledge and experience gained in other waters. We know we are reasonably skilful, know our techniques are effective because they have proved so again and again for us. So we display them in all their glory before the unappreciative western trouts, then blame the pool trout for failing to respond.

In my own early experience on English streams a No. 13 or 14 dry fly was a large one, and we commonly fished sizes down to 18. That was what the trout fell for and, so far as I knew, that was dry-fly fishing. I have taken western trout on dry flies as small as 18, and have even had one or two spectacular days with 15s and 16s. But these flies are too small.

As long ago as 1902, Theodore Gordon, pioneering with the dry fly on New York streams, noted that the insect life he was imitating was notably larger than that of the English streams, even though broadly similar in other respects. He wrote of using flies not smaller than 13 or 14 and as large as 10 or 11. As the east is to England, so the Pacific Coast is to the east; we grow them bigger, if not fancier.

My standard dry fly for general use is now usually tied on a No. 9 hook. I commonly use bivisibles on 8s and have had good results at times with bushy flies on 6s. And even 6s do not come full scale on some of the really large insects the fish take, such as the stonefly or "salmonfly" and the very large sedges.

It would be misleading to suggest that there are not hatches on western streams, and good hatches, of flies as small as those of the eastern streams of this continent and the English chalkstreams. But the first important general lesson in western dry-fly fishing is that big flies will nearly always take fish better than small flies.

Perhaps the next major point to remember is that coast streams hold both migratory and non-migratory fish. Both will come to dry flies, but they present two different problems. Broadly, the migratory fish is less predictable in every way: far less constant in feeding habits, with greater tendency to feed aggressively in midwater and a lesser inclination to take surface feed. And streams vary from savagely broken white water, rushing among great boulders, to meadow-smooth streams as quiet as can be found anywhere. The former predominate; the most useful and most specialized techniques apply chiefly to them.

I am not sure where the extreme of western dry-fly technique was developed, but I am inclined to give credit to the McKenzie River in Oregon. I say extreme; perhaps I am wrong, and maybe still more radically different techniques are used. But it is not uncommon on the McKenzie to see a man fishing two or even three dry flies, casting them across and downstream and swinging them back across the current at the end of only the briefest, drag-free float. The flies are large, on 8 or 9 hooks, winged with deer or other hair to give them good floating qualities. And the method is effective. It searches a lot of water; both finding and attracting fish that might not have been moved by a more standard method. The one obvious lesson to be learned from this is that drag, at least on big, rough streams such as the McKenzie, is by no means the hopelessly adverse factor that dry-fly fishermen usually consider it. Used wisely, it can be an asset.

My personal inclination is still to avoid drag as far as possible and to fish upstream, searching water by choice rather than by chance. I never use more than one fly because one fly is all I can, or want to, handle properly. But I no longer have the old rigid fear of drag, and very often I use it deliberately because I have found that it will sometimes make a big fish show, or even take, when the normally drifted fly will not. Such fish often come well short to the dragged fly or jump over it; when they do, it is usually possible to cover them with an orthodox fly and bring them up smoothly and confidently. I avoid fishing downstream because it brings drag very quickly and because one is likely to miss fish on the strike. But on any fair-sized stream there is much water that would be missed if one never made a downstream cast; so at times I make downstream casts and hope for the best.

These four radical departures — the large fly, fished with or without drag, upstream or down, in the expectation of an aggressive rather than an intercepting rise — add up to something like a wholly new method. They also make dry-fly fishing possible and productive where more orthodox methods are, to put it mildly, very difficult to follow. The large hair fly floats visibly and well in the roughest water. Acceptance of drag as an occasional asset rather than a certain liability gives one a whole new confidence in working water broken by many awkward currents.

The downstream method opens up a lot of water that would otherwise be unfishable. It also makes possible two other tricks of the trade that are often effective. In the first of these the fly is dragged directly upstream with a bouncing movement against broken water, then allowed to drift back on the recovered slack. In the second, which calls for a well-sunk leader, the fly is allowed to drag under, is brought up toward the fisherman, still sunk, then allowed to pop to the surface and drift back; a well-oiled deer-hair fly will do this time and again. Both methods take fish well at times.

I have taken pains so far to emphasize the unorthodox and the extreme. This is not to suggest that the perfectly cocked, perfectly floated fly, fished upstream, and matching some natural (not necessarily the main hatch) as closely as possible, will not outfish everything else under most conditions. I want only to emphasize that the dry-fly enthusiast should come to coast streams with an open mind, expecting important differences, and be ready to recognize them and to use them.

For a long while it puzzled me to find that fish preferred the large flies. Again and again I saw hatches of small, sometimes very small, May flies. Sometimes fish were rising to them, sometimes they were disregarding them. Once, on a good hatch of a small, pale May fly, I took seven two-pound, sea-run cutthroats on a No. 15 Pale Watery Dun in less than an hour's fishing. On another hatch of still smaller flies I took fifteen big sea-runs on a No. 17 Iron Blue in less than a hundred yards of river. At other times, under conditions that matched the natural so closely that I lost my own fly among them again and again, the trout were scarcely interested at all.

Long watching of many streams and the opening of many trout stomachs suggest to me what may be the answer. I have again and again watched both rainbows and cutthroats rising with apparent consistence and steadiness to some prolific hatch of naturals. But only rarely have I seen them take more than one or two out of every half-dozen of the main hatch that floats perfectly to them. What is never allowed to pass is a larger fly that happens to be hatching at the same time; the same fish that is treating the preponderant hatch so casually will go far out of his way for one of these.

The same story is in the stomachs one examines. Almost never is a coast trout feeding with complete selectivity. He will have in him, wedged among the lesser yields of the main hatch, two or three May flies, several times as big, a yellow-jacket, a big spider, a cedar-beetle — almost any variation one can imagine that is bigger and fancier.

In all this discussion I am thinking in terms of fish of twelve inches or larger, mainly in terms of fish longer than fifteen inches. There is a favourite coast habit of cradle-snatching young steelhead and salmon on their way down to salt water. These six-, eight- and ten-inch innocents will rise with eager faith to almost any dry fly or, for that matter, to the ashes knocked out of a pipe or a piece of thistle-down. Fishing for them is not sport, nor related to sport, and it is safe to assume that no true angler will be concerned with them.

At the same time it seems as well to describe in some detail the sort of streams and fish that I am thinking of. The first stream that comes into my mind is well back from the coast, at an elevation of over two thousand feet, and full of rainbow trout, averaging a pound, and running up to three pounds or more. It is a small stream this far back, formidably fast and broken, but with many good pools and long glides. In September there is already frost at night and there is little surface hatch before noon.

I am thinking now of a part of the stream that is quite accessible and has been fairly heavily fished. But remember, this means only that the fish have been consistently harried and insulted with worms and salmon eggs and spinners. The odd one may have a single-egg hook in his gullet and a nine-foot length of nylon hanging out of his mouth, but he is probably fairly unsophisticated about flies, especially floating flies. During the morning, before the hatch comes on, there will not be a sign of a rising fish anywhere, and even the La Branche method of creating an artificial hatch by repeated casting is not likely to produce much. As later examination of their stomachs shows, the fish are well down on the bottom, feeding chiefly on caddis grubs and May fly nymphs.

With the hatch they begin to show. It is probably a hatch of smallish May flies — perhaps a No. 14 hook would match them. But if you fish such a fly, it must be placed perfectly over the fish, without the slightest drag, and even then it will be disregarded more often than not. It will also be a very difficult fly to follow in the swift and broken water unless the light is just right.

Put up something like a No. 9 McKenzie Orange or Yellow Caddis, tied with a deer-hair split-wings and a red hackle palmerwise down the body, and the trout come to it with confidence. They will not only slide over a foot or two to take an inaccurate cast, but often turn suddenly and chase several feet downstream after a fly that has passed them while they were otherwise engaged. Drag is best avoided under these conditions because of the fact that it produces short and clumsy rises and the fish are showing themselves without it.

There are difficult fish in this stream, of course, and smoother places where the fish rise more deliberately and are less inclined to move out for an inaccurate cast. It may at times be a good thing to change down to a smaller fly, even to match the preponderant hatch exactly for a really difficult fish. It is important also to watch for the larger naturals that are coming down and to match the No. 9 to these. Almost any fairly large fly will probably get rises, but often many of these will be short enough to result in poorly hooked fish unless something more than the size of the fly is right.

A fly that rises fish but does not hook them securely is in many ways worse than a fly that does not move them at all. The fly that is disregarded can be changed, and the fish is still there to be covered by something else; but once a fish has been hooked and lost, he is not likely to offer a second chance. I change quickly from a fly that is getting short rises and immediately from one that loses me two fish in succession by tearing away for no apparent reason.

The right fly is one that is taken confidently on its first perfect drift over a rising fish and brings that fish, securely hooked, into the net. This sounds like an obvious point, but many fishermen overlook it and persist in using a fly to which the fish are coming short. The logical change is to something that echoes the colour of the natural hatch, and this often makes all the difference.

I remember an evening rise of cutthroats on a small mountain stream that I had fished many times before. The fish were coming freely to a small May fly, but I put up a largish grey-bodied sedge which had often done well for me under the same conditions. After several casts I had risen only two fish, both of them short. I checked the natural hatch and found that a No. 14 Blue Quill fitted it almost exactly; so I tied one on and covered half a dozen rising fish in succession without a response. Though I had been watching for them, I had seen no large naturals at all, but I knew I had to go back to a large fly; so I tried a No. 9 McKenzie Blue Upright. Every fish to which I offered it took it confidently and securely.

Colour is not always the important factor. I have often found a well-made fan-wing Coachman a perfect fly with both cutthroats and rainbows that were rising freely; fish after fish has taken it without hesitation when flies that seemed to me far closer to the natural produced only half-hearted rises. A parachute-winged fly of the Royal Coachman type has proved extraordinarily successful for season after season on a group of mountain streams where I have never seen anything remotely like it on the water; it continues to kill fish even after the parachute hackle has begun to unravel from the hump on the hook shank. Flies of the bee type, with black and yellow or brown and yellow bodies of silk or chenille, also kill consistently, but that is to be expected because of the fish picking up a certain number of drowned or drowning bees and yellow-jackets as well as deerflies in most heavily bushed streams.

These clear mountain streams are a dry-fly fisherman's paradise at their best times. They are very cold, and the fish spread up into them only late in the season when the water warms to forty-five degrees or more; but for a short distance above their outlets into lakes they are often good for several months.

The lower coast streams, immediately open to salt water, are a different problem altogether, and each has its own peculiarities; but they can be made to give the dry-fly fisherman sport, and at times a dry fly will outfish anything else. Even in the estuaries big sea-run cutthroats of two, three, and four pounds often turn to quiet feeding as the tide is flooding. A brown and white bivisible, a standard deer-hair pattern, a McGinty, or a dry brown hackle is usually exactly what these fish want. I have sometimes caught them on very small flies, 17s and 18s, but 9 to 12 are more likely sizes.

I find the really large cutthroats are often quite difficult to persuade with dry flies when they have moved well up into the rivers. Fish of fifteen or sixteen inches take well, but the twenty-inchers show at the fly without touching it, especially in the middle of the day. A change of fly sometimes brings an honest rise, but more often one has to go to the wet fly to get a reluctant fish; I do not believe this is due to the sophistication of the fish so much as to his contempt for surface feed.

Such things change from day to day, and often according to the time of day. I am always content to search a good coast river with dry fly in August and September, drifting to smoother places in the rapids, the edges of the bouncing current at the head of pools, the quiet glides where the water begins to pick up speed again at the outlet of a pool. Working in this way, I steadily learn new things that make the dry fly more effective for me. Last September I found in my own home river a run of beautiful rainbows fresh from the sea, something I had never before noticed at that time of year. Most were fish of about fifteen inches that took the deer-hair patterns readily; but I found a few of over two and a half pounds along the edges of the rapids, and these, on 3x gut, fought me like full-sized steelheads.

There are exceptions to almost everything I have written

here. Waters where fish have learned to suspect the large flies and must be worked for cautiously with small hooks; rare times when every fish seems concerned only with the preponderant hatch and a good imitation will take better than anything else. But the keen dry-fly man who keeps his eyes open and is willing to adapt his technique to what he sees can confidently expect to find sport on the coast.

In general the old dry-fly rule of fine and far off remains the best advice there is; fish upstream, searching the water ahead with a good length of line. Use deer-hair patterns on 8 to 10 hooks, and gut no heavier than 2x, preferably 3x; fish the heavy water that would be too heavy, and possibly too deep, for brown trout. Don't be too much afraid of drag except on very smooth water. Drift the fly over every likely spot, but don't waste time creating artificial hatches; it is better to move on and search more water. Be shifty and aggressive in technique. Any time a fish is attracted to the fly something is learned, and a fish that shows well short is always likely to come again to a better float. If fish are rising steadily, treat them respectfully, but remember that they are probably rising from a greater depth than brown trout. Pay attention to the natural hatch if there is one and if fish are rising to it, but don't be bound by it. Count on the big fly!

18
Freedom of Choice
(1969)

I have never, I suppose, been as enthusiastic and respectful as I should be about the theory and practice of exact imitation in fly fishing. Not that I wasn't raised properly. My angling youth was full of light and dark olives, pale wateries, iron blues, and blue uprights, and from time to time I made quite conscientious attempts to come at the nearest possible match to whatever it was the chalkstream brown trout were taking. Such attention brought its rewards. But so also, and perhaps more frequently, did departures from orthodoxy. In fact, many departures were themselves so well recognized that they took on a form of orthodoxy.

I remember a large brown fly called the Pheasant Tail that would often take a fish whose only interest of the moment seemed to be a hatch of pale wateries or dark olives. I wonder now what might have been achieved under such conditions with a good western hair-wing. I remember too that a blue-winged olive imitation was rarely as good on a blue-wing hatch as were the Ginger Quill and Orange Quill. I remember the troubles I created for certain wise old brown trout with well-tied red and blue variants. I remember how frequently, in doubt, I turned to such patterns as Tup's Indispensable and the Halfstone, with remarkably good effect.

Later on, fishing mainly for rainbows and cutthroats, often anadromous rather than brown trout, I soon became satisfied with impressionistic approximation rather than attempts at exact imitation. And examination of many trout stomachs, including those of brown trout, has long since persuaded me that precise selective feeding is the exception rather than the rule on most streams. Why should a fish refuse a good morsel of food that comes conveniently to him, just because it varies from the preponderant hatch? I do not question that there are times and places when he will. But there are many times and places where a variation from the standard diet of the day seems to be precisely what he is looking for. A fly several sizes larger and considerably lighter or darker may be taken with complete faith when a good imitation of the main hatch is refused.

However much freedom I may permit myself in the choice of patterns, I remain convinced of the importance of proper presentation. I do not, for instance, care for a dragging dry fly and I will go to considerable effort to give fish a first chance at a quiet, free float as close as possible to the natural line of creatures drifting to him. I know that a dragged fly will often raise rainbows, cutthroats, steelhead, and even some brown trout when a natural drift will not. But the chance of secure hooking is far better on a natural drift and the risk of putting a fish down is much less. With a wet fly I pay constant attention to depth, speed, direction and other aspects of motion. One knows little enough and can usually see little enough, but in time care and attention do develop an extra sense of what is appropriate to any given set of fishing conditions.

The emotional pleasures of iconoclasm can be very great. Successful use of a tiny chalkstream nymph on a wild western stream was a sharp emotional experience, though I don't think I should feel so happy about introducing a Muddler Minnow to a chalkstream. Persuading winter steelhead to take Atlantic salmon patterns and vice versa is, on the other hand, a very healthy exercise for any angler who worries too much about what pattern he is using. And the only pleasure greater than taking fish on a floating fly where the natives say they never have been and cannot be so taken is in taking fish on a fly of any kind where it is supposed to be impossible to do so.

A certain freedom from concern about precise pattern can also permit far more interesting reasons for choice than the mere preference of the fish. Fishing for Atlantic salmon in Iceland recently, I became aware that the preference of the fishermen, and seemingly of the fish also, was for small, dark-bodied flies. I have a great affection for Thunder and Lightning and had no doubt it would perform successfully — in fact I took the time to satisfy myself that it would. But I also have a feeling that of all small, dark-bodied Atlantic salmon patterns, the Blue Charm is supreme; don't ask me why — tradition, I suppose. I also remembered that I had with me a fly box that had belonged to my Uncle Decie in which were a great many Blue Charms, rather surprisingly because I had not realized the fly was a favourite of his. Since Decie had taught me to fish in the first place, I felt it would be more than appropriate to give his flies some service.

When I looked at the box, my pleasure was the greater because I found that a high proportion of the Blue Charms had been cut off the leader at some time, leaving the knot still in the eye, so I was able to be sure that I was fishing not only a fly he had owned, but one he had actually used. For the rest of my time in Iceland, those were the only wet flies I used. How many fish I took on them, I do not care to count. But it was a lot.

In one smooth, shallow, difficult little pool, on a calm and difficult afternoon, I was able to compound my pleasure by using something else that had belonged to an old fishing friend. The fish were treating me a little contemptuously, rolling at the fly and missing it. I decided that my 9/5 leader was too stiff and heavy for the No. 8 Blue Charm. Searching for something lighter, I came upon a 1x leader that had belonged to Tom Brayshaw. The combination worked immediately and I killed a twelve-pounder, not without some anxious concern for the light leader.

The concern wasn't strong enough, though, and I tried to repeat without reknotting the fly. A wild fish broke me and I had to remember that the last time I had let this happen it had been with a steelhead on the Thompson River with Tom Brayshaw standing beside me. Only that time I had retied the same knot when changing a fly instead of leaving the fly unchanged.

I looked through the box for another No. 8 Blue Charm with a knot still in the eye. I didn't find one — Uncle Decie had been more partial to 4s and 6s — but there was a Silver Blue, so I knotted that to the remains of Tommy's leader. It worked just as well and this time the leader held, though the fish was equally wild.

Perhaps it doesn't pay to be quite so capricious in the choice of trout patterns. But the next time I find myself on a good summer steelhead stream I shall be trying the effect of a No. 8 Blue Charm, fished right up under the surface film. I have little doubt of the result.

19
The Evolution of a Steelhead Fly
(1976)

Just fifty years ago I went into a tackle store on the Pacific Coast and asked if I could see some steelhead flies. The clerk shook his head in wonderment and perhaps distress. Then he hollered into the back of the store: "Say, Al, guy here wants steelhead flies. We got any?"

Al came out, obviously somewhat puzzled. "Never had any call for 'em," he said. "What kind of flies would they be?" Then he brightened. "Wait a minute, I think I got something back here." He disappeared briefly and came back with some rather dusty cards to each of which was sewn a single enormous fly. The hook size was about 2/0, as I recall it, but the irons were heavy enough to hold a shark. The most prominent pattern was a Red Ibis with a gleaming scarlet body at least three eighths of an inch thick and a dyed swan wing almost large enough to lift the original owner. There was a Coachman of similarly opulent proportions, a trout pattern Silver Doctor and probably a Royal Coachman, though I am not sure of this.

I bought them, of course — who could resist? — but I don't think I recall ever putting one in the water. If I ever did swim one, it was the Coachman, in Chile, at the request of a guide who saw it in my box. It was immediately pursued and taken by a very large rainbow. Glancing back over my boxes I am now afraid this was a much lighter fly, tied on an ordinary No. 1 hook, and acquired at some other time; I have been unable to find any survivors from that 1926 purchase.

After this first introduction I gave up asking about steelhead flies and simply used standard Atlantic salmon patterns quite successfully for summer fish, usually in sizes from No. 2 to No. 6; at the same time I noticed that summer fish were also likely to take ordinary trout patterns such as Teal and Silver and my own Silver Brown on hooks as small as 8 and 10. In those days I rarely fished with a fly for winter steelhead.

The first fly I ever saw that was specifically tied for steelhead was the General Money Special. It was an effective fly and I gave the dressing in my book *The Western Angler* as follows:

Body: Black silk
Ribs and tag: Gold tinsel
Wings: Red swan with or without topping over
Hackle: Yellow

This was taken directly from a fly tied by the general himself. I have seen a good many variations since, adding a tail, substituting wood or seal's fur for silk in the body or ribbing with silver instead of gold tinsel. General Money also showed me his Prawn Fly which I found extremely good for winter fish. Again the original dressing is very simple:

Ribs and tag: Silver tinsel
Body: Orange or fiery orange wool

A long, red hackle wound palmerwise along the body. In most winter conditions a No. 1 or 2 hook seemed about right.

At about this time, somewhere in the early 1930s, I began to learn of good steelhead patterns from the State of Washington through my friends Letcher Lambuth and Enos Bradner. The Skykomish Sunrise and the Purple Peril were among them, and such California favourites as the Golden Demon.

I think it was not until some years later that I learned of other Oregon and California patterns — Harger's Orange, the Comet, the Boss Fly, the Umpqua Special, and the truly remarkable Skunk Fly, among others.

I am not myself a good or even very enthusiastic fly tyer. I usually turn to the vise and feathers out of irritation, frustration, or puzzlement, though I sometimes have a bright idea that will not be denied. During the 1930s I made a serious effort to develop reasonable representations of the several species of Pacific salmon fry. One of the most successful of these attempts resulted in the Silver Brown, an imitation of coho salmon fry that proved really attractive to sea-run cutthroats in late summer and fall. I made a slender, lightly dressed fly on No. 6 or No. 8 low-water hooks, using the full length of the hook, as follows:

Tail: Small whole breast feather of Indian crow
Body: Flat silver tinsel
Hackle: Dark red game cock
Wing: Slender strips of golden pheasant centre tail enclosing a few strands of orange bear fur

Indian crow is hard to get now, but it makes an excellent imitation of coho fry's orange tail. A very small whole tippet serves the same purpose, probably just as well. Even for cutthroats I fished this fly very slowly, just under the surface film, and since I was experimenting with greased line techniques for summer steelhead at that time, it wasn't very long before I gave them a chance at the Silver Brown. The steelhead were delighted with it. It is a fly I still use with the greatest confidence from July to October. Besides cutthroat and summer steelhead, it has taken brown trout, Atlantic salmon, eastern brook trout, Dolly Vardens, and coho salmon in both fresh and salt water.

During the Second World War, when I was away from steelhead fishing, I spent quite a little time thinking of the ideal winter steelhead pattern. I was already getting the idea that the pattern itself did not matter nearly as much as size and type and presentation, but I still felt it would be logical to follow the general consensus as to the winter steelhead's preference for red and orange. I found my thoughts returning again and again to full-dressed Atlantic salmon patterns like the Durham Ranger, Red Ranger, William Rufus, President and, above all, the Red Sandy which decorates the spine of the jacket of the Lonsdale Library *Salmon Fishing*. All have one point in common: large golden pheasant tippets in the wing. They are also complicated and beautiful dressings, which wouldn't do for me at all, so I simplified down to what I felt were the essentials and ended up with a pattern I called the Golden Girl.

Tag: Orange tying silk
Tail: Two small toppings
Body: Flat or oval gold tinsel, built up to about twice the thickness of the hook shank
Hackle: Yellow
Wing: Two large tippets enclosing orange bear fur, with a topping over all

In the years immediately after the war I used this fly a great deal for winter steelhead, usually on standard hooks from No. 2 to 2/0. The Golden Girl took fish very well and I noticed I could often feel a fish gently touching the fly on its swing and then, by careful manipulation, persuade him to take as it straightened out below me. Soon I noticed that I was losing a lot of tippets and at first I thought they broke off in casting; in time I realized that a fish following on the swing often plucked them off. This materially reduced my satisfaction with the fly. I still use the Golden Girl, but a fly that I believe is called the Fall Favourite makes more practical sense. The dressing is about as simple as it can be:

Tail: Red
Body: Oval silver tinsel
Hackle: Red, long enough to reach to the barb of the hook
Wings: Orange red bear fur

Hook sizes 1 to 4 seem about right for this fly and I think it offers all the essentials of the red-orange shrimp class of flies that are found in every steelheader's box. I would add that I think a fly of this general type and colour makes very good sense, since both salmon and steelhead often feed extensively on reddish shrimps during their saltwater lives.

In fishing for anadromous fish in fresh water, where they are feeding very little, if at all, one is attempting to trip the mechanism of a dormant feeding reaction. I believe this mechanism may be based either in the saltwater years or the juvenile freshwater years. In the summer months the early freshwater years are probably more influential, and this would explain why small, dark flies are usually so successful for summer-run fish, both steelhead and Atlantic salmon.

Shortly after the Second World War I began to notice that floating flies commonly produced more action from summer steelhead than wet flies. I had long used the old McKenzie bucktail floaters for cutthroats in fast broken water because they floated well, and these were the first floaters that rose steelhead for me, so far as I recall. I had a fair proportion of short rises, missed rises, swirls, lunges, tail slaps, and follow-backs, and this led me into search and experiment over several years to find something more convincing. I won't go into detail except to say that some of my worst looking flies, creatures that flopped over on their sides instead of riding upright, produced the liveliest action and the most honest rises.

This convinced me that what I needed was a representation of a terrestrial insect, drowned or drowning. I wanted a relatively dark colour, tending to brown rather than grey, hair wings and tail, because of its floating qualities, a fairly heavy body that would settle down into the surface film, and a very sparse hackle. I preferred a brown and yellow body because it had always done well for me and it obviously suggests a number of summer creatures that may fall into the water — bees, yellow-jackets, deerflies, gadflies, even some dragonflies. The fly that grew out of this extensive trial and error was the Steelhead Bee. It can be tied on a wide range of hook sizes, from No. 2 to No. 10, depending on the state of the water, but sizes 6 or 8 are usually about right.

Tail: Fox squirrel, quite bushy
Body: Equal sections dark brown, yellow and dark brown silk
Hackle: Natural brown, sparse
Wings: Fox squirrel, quite bushy, tied forward, divided and straightened back within about 10° of upright

This is a very effective fly, for me and for many other fishermen who have tried it. Obviously there is nothing very original about it. In type it is a Wulff, and the dressing has clear echoes of the Western Bee and the McKenzie Brown and Yellow Bug. It raises summer steelhead at least as well as any fly I have tried and hooks a higher proportion of them.

It is also a long, long way from the Red Ibis.

20
Articles of Faith for Good Anglers
(1960)

Some twenty million angling licences a year are sold on the North American continent and considerably more than twenty million people go fishing each year. There isn't a reason in the world to suppose that twenty million people really enjoy going fishing; a remarkably high proportion of them contribute vastly to the discomfort of others while finding little joy in the sport for themselves. This is sad but inevitable; it grows directly out of the misconception that anyone with two hands, a hook, and a pole, is equipped to go fishing. After all, the beloved fable has it that the boy with the worm on a bent pin always does far better than the master angler with his flies and intricate gear, so it follows logically that a state of blissful ignorance, combined with youthful clumsiness, is the perfect formula for success. If the formula doesn't prove itself, the trouble is probably the weather.

Fishing is not really a simpleton's sport. It is a sport with a long history, an intricate tradition, and a great literature. These things have not grown by accident. They have developed by the devotion of sensitive and intelligent men and they make not only a foundation for rich and satisfying experience but the charter of a brotherhood that reaches around the world and through both hemispheres.

It is a brotherhood well worth joining. There are no papers to sign, no fees to pay, no formal initiation rites. All that is required is some little understanding of the sport itself and a decent respect for the several essentials that make it.

The first purpose of going fishing is to catch fish. But right there the angler separates himself from the meat fisherman and begins to set conditions. He fishes with a rod and line and hook — not with nets or traps or dynamite. From this point on, man being man, further refinements grow naturally and the sport develops. The fisherman is seeking to catch fish on his own terms, terms that will yield him the greatest sense of achievement and the closest identification with his quarry.

This establishes the first unwritten article of the brotherhood. Fishing is a sport, a matter of intimate concern only to fish and fisherman; it is not a competition between man and man. The man's aim is to solve by his own wits and skill the unreasoning reaction of the fish, always within the limits of his self-imposed conditions. Besides this, any sort of outshining one's fellow man becomes completely trivial. The fisherman is his own referee, umpire, steward, and sole judge of his performance. Completely alone, by remote lake or virgin stream, he remains bound by his private conditions and the vagaries of fish and weather. Within those conditions, he may bring all his ingenuity to bear, but if he departs from them or betrays them, though only God and the fish are his witnesses, he inevitably reduces his reward.

This total freedom from competitive pressure leads the fisherman directly to the three articles of faith that really govern the brotherhood: respect for the fish, respect for the fish's living space, and respect for other fishermen. All three are interrelated and, under the crowded conditions of today's fishing waters, all three are equally important.

Respect for other fishermen is simply a matter of common courtesy and reasonably good manners. The more crowded the waters the more necessary manners become and the more thoroughly they are forgotten. The rule can be expressed in a single golden-rule phrase: "Give the other guy the kind of break you would like to get for yourself." Don't crowd him, don't block him, don't push him. If he is working upstream, don't cut in above him; if he is working downstream, don't pile in directly below him. If you see he is hooking fish along some favourite weed bed, don't force your boat in beside him and spoil it for both. Don't park all day in what you think is a favoured spot so that no one else can get near it — give it a fair try and move on.

On uncrowded waters a self-respecting fisherman always gives the other fellow first chance through the pool or the drift; as often as not the second time through is just as good. On crowded waters give whatever room and show whatever consideration you can and still wet a line; better still, try somewhere else. The crowds are usually in the wrong places anyway.

If you would be part of the brotherhood, be generous. Don't hide the successful fly or lure or bait; explain every last detail of it and give or lend a sample if you can. Show the next man along where you moved and missed the big one, make him aware of whatever little secret you may have of the river's pools or the lake's shoals or the sea's tides — but only if the other guy wants it. If he doesn't, be generous still and keep quiet. If he wants to tell you his secret instead of listening to yours, reach for your ultimate generosity and hear him out as long as you can stand it. Good things sometimes come from unlikely sources.

Respect for the fish is the real base of the whole business. He is not an enemy, merely an adversary, and without him and his progeny there can be no sport. Whatever his type and species, he has certain qualities that make for sport and he must be given a chance to show them to best advantage. He is entitled to the consideration of the lightest gear and the subtlest method the angler can use with a reasonable chance of success. Trout deserve to be caught on the fly; other methods may be necessary at times, but it is difficult to believe they give much joy to the fisherman. A northern pike or a muskie taken by casting is worth half a dozen taken by trolling. A black bass tempted to the surface is a far greater thrill than one hooked in the depths; an Atlantic salmon or summer steelhead risen to a floating fly is a memory that will live forever. If it takes a little time to learn such skills, there is no doubt the fish is worthy of them. And if the angler is any kind of a man he is unlikely to be satisfied with less.

Even in the moment of success and triumph, when the hooked fish is safely brought to beach or net, he is still entitled to respect and consideration: to quick and merciful death if he is wanted, to swift and gentle release if he is not. Killing fish is not difficult — a sharp rap on the back of the head settles most species. Releasing fish is a little, but only a little, more complicated. Fly-caught trout of moderate size are easy. Slide the hand down the leader with the fish still in the water, grip the shank of the hook, and twist sharply. Where it is necessary to handle the fish, a thumb and finger grip on the lower jaw does the least harm and is usually effective. If not, use dry hands and a light but firm grip on the body just forward of the dorsal fin. Wet hands force a heavier grip which is extremely likely to injure vital organs. For heavily toothed fish like northern pike and muskies many fishermen use a grip on the eye sockets or the gill-covers. The first may be all right, but seems dangerous and unnecessarily cruel. The second is destructive. Fish up to ten pounds or so can be gripped securely on the body just behind the gill-covers and should not be harmed.

Larger fish that have fought hard are often in distress when released and need to be nursed in the water until they can swim away on their own. Generally little more is needed than to hold them on an even keel, facing upstream, while they take a few gulps of water through their gills. If they lack the strength for this, draw them gently back and forth through the water so that the gills will be forced to work; all but the most exhausted fish will recover under this treatment and swim smoothly away. Fish that have bled heavily or fish that have just swum in from salt water are less likely to recover and should be kept.

Respect for the fish's living space should be comprehensive. It includes the water, the bed of the stream or lake, the land on both sides of the water, and all the life that grows there, bird or mammal, plant or fish or insect. There isn't an excuse in the world for litter-leavers, tree-carvers, brush-cutters, flower-pickers, nest-robbers, or any other self-centred vandals on fishing waters. The fisherman comes at best to do some damage — to the fish — and the best he can do is keep it to that. He doesn't need to junk-heap the place with cartons and bottles and tin cans; he need not drop even so much as a leader case or cigarette pack; he can afford to remember that no one else wants to be reminded of him by his leavings.

These are elementary and negative points and if parents raised their children properly there would be no need to mention them in this context. A fisherman, any kind of a fisherman, should know better than to spoil the place that makes his sport. But a true share in the brotherhood calls for a little more. The fisherman is under obligation to learn and understand something about the life of his fish and the conditions it needs, if only so that he can take his little part in helping to protect them.

All fish need clean waters and all nations, if they know what is good for them, can afford to keep their waters clean. Pollution, whether from sewage or industrial wastes, starts as a little thing scarcely noticed and goes on to destroy all the life of the waters. Its damage can be repaired, slowly, painfully, expensively, but there is no excuse for it in the first place, though many are forthcoming.

Besides clean water for their own lives and the many living things they depend on, fish need special conditions for spawning and hatching and rearing. Migratory fish need free passage upstream and down. These things and many others like them are worth understanding not merely because they suggest protections and improvements, but because knowledge of them brings the fisherman closer to the identification he seeks, makes him more truly a part of the world he is trying to share.

The old days and the old ways, when every stream was full of fish and empty of people, are long gone. They weren't as good as they sounded anyway. It took time and the efforts of good fishermen to learn what could be done and should be done to produce the best possible sport. North American angling has now come close to full development. No one is going to get what he should from the sport by simply buying some gear and going out on the water, nor can he achieve very much by sneering at better men than himself who do take the trouble to learn the delicate skills of the subtler methods. The real world of fishing is open to anyone, through the literature and the generosity of the brotherhood. Once entered upon, the possibilities are limitless. But even the casual, occasional fisherman owes the sport some measure of understanding — enough, shall we say, to protect himself and others from the waste and aggravations of discourtesy and bad manners that are so often based on ignorance.

In Winchester Cathedral, not far from a famous trout stream in Hampshire, England, is the tomb of William of Wykeham, a great fourteenth-century bishop and statesman who left a motto to a school he founded: "Manners makyth man." Within the same cathedral lie the bones of our father, Izaak Walton, who remarked three hundred years ago: "Angling is somewhat like poetry, men are to be born so." Perhaps Izaak's precept is for the inner circle of the brotherhood, but William's is certainly universal. It is just possible that nice guys don't catch the most fish. But they find far more pleasure in those they do get.

21
Outdoor Ethics
(1964)

Outdoor ethics, sportmen's ethics, woodsmen's ethics, call them what you will, have existed since the beginning of human time. I think, first of all, of the ethic of the cache and the campsite, the ethic of the unlocked cabin and the unlocked boat. These were, and are, the ethics of survival in the wilderness. The cache is sacred, except in time of need. Even then what is taken must be replaced, or the owner warned of its reduction. If not, he may depend on it and die for lack of it.

The mountain campsite is used, or not, for few people pass that way. But when used it is left "improved" with fresh wood cut and new trees ringed to dry out and replace others that have been used. The same is true of the unlocked cabin and its contents. The unguarded boat is there also for the stranger's use in case of need. But it must be replaced, exactly, beyond danger from water or storm, with paddles or oars or pole exactly as they were before. If not, a man may die for the error of depending upon them and upon the ethical behaviour of his fellow man.

These are simple, straightforward ethics conditioned by life and death and the hope of mutual survival. They are observed today — at least in areas beyond the roads and beyond the reach of mechanical transportation. Basically they are matters of common human decency. Where they are not observed it is because the reasons for them are not understood, because the reasons underlying them have not reached in to touch the imagination of those who fail. Imagination is a necessary factor in all ethical behaviour. Expressed very simply, and far too simply, it takes the form: "What if someone else did this to me?"

The sort of outdoor ethics I am concerned with today are vastly more complicated than those I have just described. Yet they are based on the same common human decencies and on the need for survival. In this case, the survival of outdoor sportsmen's interests, the survival of fish and wildlife, the survival of land and water in useful form, and the survival of ourselves as a people. I do not propose, in this short space, to make any attempt to offer you a code of outdoor ethics. The best I can hope to do is outline the sort of thinking that produces sound ethical behaviour and perhaps sound ethical codes.

But let me for a start examine our desperate need for ethical performance. It grows, of course, from increasing population and decreasing resources. But this is not all. It grows also from increasing wealth, increasing mobility, and increasing leisure — the last of which has not yet begun to make itself felt in anything approaching full proportion. The increase of leisure that is to be expected from growing automation will call for whole new theories of government — new concepts of society and economics and a vast strengthening of ethics in every field. I call your attention to one field where it is already in an urgent state: the automobile. We all abandon ethics to some extent when we get behind the wheel of an automobile. But when we do we are in danger. Men die on the roads every day for their own want of ethics, or for someone else's want of ethics, and for no other reason.

André Malraux is one of the most brilliant minds in the world, and as minister of culture in France is deeply concerned about this prospect of increasing leisure. Man, once his basic needs are satisfied, or so M. Malraux believes, if I understand him correctly, is a creature of blood and sex. Given a sudden excess of leisure he will turn in upon his own inner darkness.

It so happens that I do not agree with this particular assessment of the human race. Nor, for reasons that I will give later, do I anticipate the enormous excess of leisure that is suggested. But I am satisfied that we are, usefully I hope, faced with a great increase of leisure. I am satisfied that we owe it to mankind to recognize this and to do a lot about it. I believe, and have long believed, that on this continent, the outdoor resources must play a big part in caring for man's new need. I believe, also, that sound ethics in use and management will determine how well, or how poorly, this part is played.

I have met and talked to outdoor groups from Seattle, Washington, to Sheet Harbour, Nova Scotia; from Vancouver Island to Boston, Massachusetts; from Lac La Ronge, in the north, to Bariloche and Valparaiso in the south. In this kind of group there is always one ethic, universally accepted, and that is the conservation ethic — concern for the future. I think you will agree with me that that is so. Stewart Udall, Secretary of the U.S. Department of the Interior, has defined the conservation ethic in the following words: "It measures the progress of any generation in terms of the heritage it bequeaths to its successors." This, I submit, is a sound and positive statement that stands up to a lot of examination.

In the present context it means a balance sheet of management, use, restoration, and protection that always totals to a significant advance upon the past. Some values are to be used and improved. For instance, stocks of fish and wildlife and their habitat. Some values must be preserved inviolate, by firmly controlled use. The national parks belong in this group. So do wildlife sanctuaries. Values destroyed or damaged must be compensated or repaired.

Now a positive and progressive balance sheet of this sort is not maintained by hasty or careless development, by sudden new policies or without constant effort. On that point I commend to you some words of the late President Kennedy: "Each generation must deal anew with the raider, with the scramble to use public resources for private profit, with the tendency to prefer short-run profits to long-run necessities. The nation's battle to preserve the common estate is far from won. We must do in our own day what Theodore Roosevelt did sixty years ago, and Franklin Roosevelt, thirty years ago. We must expand the concept of conservation to meet the imperious problems of the new age."

You will notice that I have been at some pains not to quote "woolly-minded conservationists." I have quoted France's Minister of Culture. I have quoted the U.S. Secretary of the Interior. I have quoted the late President of the United States. If you can write them off, do so.

I suggest, though, that we are in excellent position to do exactly what Udall and Kennedy suggest. We are beginning to understand the population dynamics and habitat needs of our wildlife stocks. We have a new and constructive understanding of the spawning, hatching, and rearing needs of our enormously important fish stocks, and let me assure you that there is barely a stream on the continent that cannot be materially improved in these respects. We have the experience and guiding development and philosophy of several generations to direct the use and management of our park areas, if only we have the wits and the humility to draw upon it. We can have no possible excuse for failure in the conservation ethic.

Let me turn again for a moment to automation and its threat of excessive leisure. I refuse to believe that there can be such a threat while so immense a proportion of the world's work is not being done: streams are left unimproved, parks undeveloped, forest and grazing land uncared for — to offer only a few examples in our special field. When the push-button economy arrives, I cannot, and do not, believe that the profits will be left with the inventors of buttons or the owners of buttons, or even the pushers of buttons. I believe, rather, that they can and will return to the performance of these essential duties that are left undone today.

So far I have discussed ethics on the broad plane — now let me return to the more personal plane. I submit first of all that there is no such thing as sport without ethics. The angler uses rod and line and gear more or less carefully calculated to his fish; the shot-gunners kill their birds on the wing; the big-game hunter selects his quarry and terrain with care, and looks for challenge to his woodcraft. Without these sorts of ethics there is not any sport.

From these traditional ethics a more complicated system of ethics grows and builds. And behind all such traditions there is one guiding spirit — attitude of mind, and it has three major directions. One, consideration for his fellow sportsmen. Two, consideration for fish and wildlife stocks. Three, consideration for the land and water upon which they and his sport depend. These are things that have to be taught from the very beginning. If they are taught and properly understood, more specific codes of ethics will inevitably grow from them.

Ethics, though they accept the law and abide by it, are, as I said at the beginning, a morality beyond demands of the law. You cannot legislate against the hunter who misbehaves and impairs the access rights of other hunters. You have to train him. You cannot legislate successfully against the jealous fisherman who spoils the sport of others as well as his own. But, you can teach the spirit of generosity; you can establish codes and standards — written or unwritten — that will, in time, find general acceptance and attain a moral force more powerful than the law.

You cannot do this in a week or a year, but it can be done over a period of years, if you, as leaders, tighten your own standards, define them more carefully to yourselves, and pass the results of that experience on to others. This is precisely what will be needed as more and more people go out into the woods and onto the waters to find satisfaction in their own particular way.

I have written and spoken more specifically elsewhere of what I consider ethical standards for hunters and fishermen. But I should like to describe, very briefly, a more recent problem in outdoor ethics that faces me personally. In my old age I have turned, with a most immoderate enthusiasm, to the sport of skin diving. This is a sport that has a very high code of ethics of its own, especially in regard to safety and consideration of one's fellows. But I had one or two other matters to resolve. The first promise that I made to myself in taking up skin diving was that I would not kill or unnecessarily molest anything under water — believing that peace between man under water and other creatures under water is a state that should be preserved. I faced, also, the possibility that I might learn under water something that would unduly improve my fishing techniques. But after a little thought, I decided that this was a fairly remote hazard. Then, with these ethical considerations disposed of, I set out. I need hardly tell you that my main interest has been in the rivers and in the salmon and trout that frequent them. Watching the various salmon runs come home last fall was one of the supreme satisfactions of my life. But it immediately raised another problem in outdoor ethics.

There was no possible doubt in my mind that my presence frequently disturbed the salmon and sometimes scared them into a frantic flight that I knew must be drawing, to some extent, upon reserves of energy they needed for spawning. By close observation and an involved system of reasoning, which I won't go into here, I decided that, so long as I exercised what skill I could and what caution I could, the stresses caused by my disturbance of their affairs would not be unduly high and that under these conditions I was entitled to go on observing them. But, I must add to that, if diving to observe salmon in streams ever becomes a popular sport, with large numbers of divers in the rivers, this conclusion would have to be entirely different. I should no longer feel justified in adding to the disturbance of the fish, and ethics would keep me from doing so, even if legislation did not.

I think this is the way in which the ethics of any sport or activity must be worked out — on a personal level at first, and later, if necessary, on a general basis. The ethical standards must be constantly strengthened and improved in all forms of outdoor activities, I have no possible doubt; that laymen must play the major part in this, and professionals must listen and understand, I am equally certain. If use of the outdoors, the woods, fields, the lakes and the streams, is to continue a harmonious, productive, and satisfying activity for increasing numbers of people, we're all going to have to behave much better than we do today. We will need increasing consideration for others and a lively spirit of generosity in all our doings.

Public ownership of the outdoor resources is a tremendous co-operative experiment. We shall always have to fight to keep it that way. A strong code of outdoor ethics, well considered, faithfully observed to the point at which it becomes a tradition, will be no small part of the outdoor heritage that we can bequeath to our successors.

22
A Talk to
Oregon Fly Fishermen
(1961)

Mr. President, ladies and gentlemen: I feel I should first congratulate you for achieving such a very large membership in such a very short time. I am sure the waiting-list is going to be long and unhappy before very long. I do congratulate you on your venture and I do wish you every kind of good fortune because I feel that fly-fishing clubs all over the United States and Canada have a great deal to do for fishing generally and for themselves in particular.

We have had troubles on this coast as fly fishermen. I have been here for thirty-five years. To hear people talk, when I first came here, you would think that no western trout really took a fly in any serious way at all. It was a chancy matter, something like the old story that you couldn't sail a boat in Puget Sound. Long Island Sound was all right, but Puget Sound wasn't. You just couldn't sail in Puget Sound. It was rather the same about the Pacific Coast trout. You couldn't catch them on a fly. You certainly couldn't catch steelhead on anything but salmon eggs. I think and hope we're growing out of that a little.

I remember to this day (sometimes to my blushing shame when I am in my more respectable moments) that the first way I learned to hunt ruffed grouse was with a good tree dog and a .22 rifle — just go out in the crab apple swamp and let the dog loose; when she barked there was a grouse up a tree, and you shot its head off. I thought that was a great deal of fun. But I assure you I wouldn't do it today, and for the same reasons I hope we are growing up out of some of the more pragmatic attitudes of the pioneers towards the techniques of taking fish.

I didn't fish salmon roe then because I had been better trained as a fisherman than as a hunter. When we really needed fish for the winter, I preferred to put out a short length of gill net and take them that way than to use salmon roe. I still would. But the funny thing about it is that even in those early days, while there were a lot of good fly fishermen here, they were not very vocal. You had a job to find or meet any. But if you could fish a fly competently, it was surprisingly easy to convince the pioneers that this technique had something. He would be out there with his pragmatic methods, and if conditions were right, you, with your mysterious waving of the strange pole around in the air, and the kind of tackle you had brought with you, could vastly outdo him. Many of the ones I met rapidly converted to these more subtle methods. I think this goes on. Fly fishing is, in the last analysis, quite an effective method of fishing. This is the point that has to get over to the public to a greater extent than it does today. We have had a lot of very good and very ingenious fly fishermen working here now for a long time. I think we can say the West has come into its own.

The West has always had one of, if not the finest selections of fly fish to be found naturally anywhere in the world. There is no other part of the world that has two native true trouts, one of them a sea-running fish so similar to the Atlantic salmon that you really can't measure any great difference. All the potential of the Pacific salmon on the fly, plus the Dolly Vardens, which are after all prevalent in Europe, too, and known as the Arctic char. Plus, in some parts of our West, the Arctic graylings and the Montana grayling. And I haven't mentioned the whitefish, which is no contemptible creature, whether he is the Great Lakes whitefish or the Rocky Mountain whitefish. Either of them is worth catching on the fly. That's quite a list of first-class fly fish, and we now have the tackle and the techniques to meet them on their own grounds. I don't see why we can't be just as good as anybody else has ever been at this business — perhaps a little better, and a little more versatile with a fly.

At the same time a little humility is probably in order because we are simply extending a tradition and a brotherhood that already almost meets around the world, and reaches into both hemispheres, as you know. You remember how this tradition grew first in Britain and how it has been reflected there in literature spanning almost five hundred years. It is about five hundred years since Dame Juliana Berners first produced her great book, followed by Gervase Markham. Then you go on to Walton and Cotton; and then a big jump to Ronalds and Stewart, and finally down to Halford and Skues, all on the other side of the Atlantic. That is the hard line of the tradition. There are many, many other admirable lights in amongst these. The tradition crossed the Atlantic with Frank Forrester. Later came men like Theodore Gordon who developed their own methods and their own almost mystic or instinctive fly fishing techniques. These were highly sophisticated, highly intelligent adaptions of European methods to eastern streams. And so the thing grew through our other great names in the East, such as Hewitt and La Branche. In Canada, though they were less technicians but perhaps better writers, there were Blake and Comeau, both first-rate performers.

It still takes, however, ordinary fly fishers like you and me to do the real work and make the thing stick. It is a little bit of a burden to carry, this great tradition of fly fishing, but it is one I think we should welcome and accept gracefully, because as a technique and philosophy it has produced so well in the past, and it can go on producing in the future. If we slight it, it won't.

I think fishermen have always been in the habit of forming themselves into either semi-formal or extremely informal clubs. I believe fly fishing as such started in Macedonia somewhere around the fourth century. I would rather suspect that the first fishing clubs were formed in the British pubs, obviously a natural nursery for such affairs. If you remember your Walton, the old gentleman, while a very religious type and a little bit of a Puritan in his ways, was always anxious or ready to knock off on the stream and go to the nearest pub for a cheerful glass and a round of song. I'm delighted that he did so. Then, of course, the real patron saint of fly fishing and fly tying, Charles Cotton, was a most delightful rake. He not only wrote his share of the *Compleat Angler*, but was also an excellent poet — one of the less bawdy of the Restoration poets, but still bawdy enough. He was probably one of the best translators of Montaigne's essays and a distinguished diplomat in the Court of King Charles II. Cotton was a very merry individual, and this belongs definitely in our tradition. In any event, anglers should be cheerful people. We should take delight in all the pleasures of life.

I've really lived too much in the wilds to have known a great number of clubs, but I've always been impressed by the particularly delightful nature of the clubs I have known. I am sure that this would be true of the Fly Fishers' Club of London, an establishment I have never visited. However, it does overawe some people. My uncle Decimus (he was called Decimus because he was my grandfather's tenth son) is a very good fisherman. He taught me to fish, and he became a member of the Fly Fishers' Club of London. I once asked him many years after he had joined that club how he liked it.

"Well," he said, "as a matter of fact I've been a member for twenty-five years, but I've never dared go in the door."

"Why not?" I asked.

"Well, I think there would be an awful lot of stuffy old gentlemen in there, and the first thing they would do would be to get me into an argument I couldn't finish."

I think since that time he has had other thoughts, and I

think he has gone inside the club and enjoyed it very much. I can assure you if I were a member of the Fly Fishers' Club of London I wouldn't wait twenty-five years to go in, much as I respect my Uncle Decimus.

Now the New York Anglers' Club, I do know. And I can assure you that they are a very relaxed, a very friendly and pleasant group indeed. They have built themselves a handsome tradition in a very short time. They have many pieces of club property that would be of interest to anybody in the fly-fishing world, including such things as Theodore Gordon's fly boxes. It is a fascinating place to visit, quite apart from the delightful nature of the people you meet. The club's possessions make what is really the "holy of holies" of angling on the North American continent.

Another group I know well is the Boston Fly Casters. An extremely informal and very merry group. I don't think they have any club quarters of their own or any burdensome traditions, or anything else but a very good time. I always have too good a time when I get there.

You all know the Washington Fly Fishing Club, I imagine. If you don't, again, here is a very cheerful, pleasant and informal group. And this time, of course, a western group. As far as I know, the Washington Fly Fishing Club just grew naturally.

In British Columbia we have an informal group that you might be interested in hearing about. It's called the Hawthorn Foundation. A number of us, including the president of the university and his assistant, and others of the University of British Columbia, are in the habit of meeting for a fishing trip right after graduation. And we usually have a little pot on the first fish, the largest fish, and so forth. We also play a little poker in the evening. Stakes are not high, but one year one member seemed to be on the winning edge of things — excessively so, I may say. As it happened, I told him and another very reputable professor, about a little lake where there were a lot of good-sized fish that you could catch very readily. These two went out the next day to this little lake, and the second of the professors (the one who had all the winnings) became very excited. He stripped to his shorts, got on a log, and paddled out into the middle of the lake. He started murdering fish right and left. His friend on the bank kept fairly quiet, taking a fish now and then. When they came back to the lodge where we were staying somebody met Stan (who was the quiet man on the bank) as soon as he came in and said, "Stan, how many fish did you get?" Stan said, "Eight."

Well, our limit up there is twelve, and when the fish were counted there were twenty-one. It was quite clear that the law had been broken, and it was also quite clear who had broken the law. So we held court that evening, and decided that all the winnings of this character who had taken thirteen fish instead of twelve should be placed in escrow until we decided that the winnings would be given to the university library to start a fishing section called, after this gentleman, the Hawthorn Foundation, "for the promotion and inculcation of the ethics of fly fishing." This club meets every year, and the poker winnings, no matter who wins them, are still forfeit. So are any other ill-gotten gains that anyone may have, in addition to which, of course, it isn't too hard to penalize members from time to time for various infractions on the ethics of fly fishing. We are building up a nice little library in the University of British Columbia on the subject of fly fishing. That is one way a club can develop and serve.

Another example of the sort of thing a club can do that has interested me recently had to do with developing quality, rather than quantity in fishing. A club in Vermont that owns some of its own water wrote me not long ago to say that they had decided to change their stocking methods, and control and adjust all the management methods of their lakes to produce quality rather than quantity fishing. This is, to me, the essence of fishing. It's just as applicable under public ownership conditions as it is under private ownership conditions. We are just as much entitled to ask for quality fishing, even though we may be in a minority, as is a private club to vote itself into quality fishing rather than quantity fishing. We have vast public waters here. We have increasing pressure on them all the time. We cannot expect forever to go out and just take fish out of them, particularly salmonoid fishes, by any methods at all in large-size limits with plenty of room for everyone. It just doesn't happen that way. For this reason alone, I think we have a right to ask for quality fishing. It doesn't necessarily mean fly fishing-only waters, but it probably means quite a lot of fly fishing-only waters in many parts of your state and my province, as well as in other states and other provinces.

In British Columbia we are trying to work for fly fishing-only waters. We are not pushing the thing terribly hard, but are simply bringing the matter up again and again. Gradually the authorities begin to understand what we are talking about. We undoubtedly will get a number of fly fishing-only waters in time. The point is that if you do have public waters, you do have biologists and game management people running them. They are not only managing for the fish, they are managing for the people as well, and minorities have rights as well as majorities. Fly fishermen are a very strong minority and their rights should be respected. There is no overall inherent right to fish in every water in Oregon, Washington, California, and British Columbia, with worms and salmon eggs just because some people say they want to. There are places for that, and places for the fly, too. Remember, they do this in the East. In New Brunswick, Nova Scotia, Labrador, Newfoundland, we have fly-only regulations for Atlantic salmon, and nobody complains of any great hardship there. If they can do it for Atlantic salmon, heaven knows we can do it for trout on many of the streams. I think our fish are of sufficient quality to rate the respect of fly-only regulations.

In this context it is important to remember that the fly fisherman's sport is in many things besides the catching of fish. We pride ourselves in knowing something of our fish, how they live and behave, what water conditions they need for living and breeding, and how and when and where they feed. The insect life of lakes and streams has always been a special part of a fly fisherman's study. It is satisfying to be able to recognize and name waterside trees and shrubs, to trace with a knowing eye the passage of the previous winter's floods, watch the build up of a gravel bar from season to season or the scouring of a new pool.

Waterside birds can make a large contribution to a day on a stream or a lake. I think we should be able to recognize most of them without difficulty and am always astonished because so many fishermen cannot. Heron, water ouzel, merganser, spotted sandpiper, osprey, kingfisher, and many others are all part of going fishing and it is a great pity not to know them.

A good part of the pleasure of going fishing is in understanding these things, watching them and recording them in the mind, being able to name them and hold them for yourselves as valued things. Identifying them and knowing something about them gives you a special claim in your own world of the water's edge, and helps to make you a part of it instead of a mere intruder. This to me is a very important thing. It gives a sense of identification with the whole natural world which I think most of us are looking for. As nearly as I can find any one reason for why we go out to hunt and fish it is in this search for a sense of identification with the natural world. No one finds it more completely or more rewardingly than the fly fisherman. Yet in searching for it, we have no need to damage or reduce anything of this precious environment. If we understand our part, we can pass through that world with as little trace of ourselves and our passing as the Indian left when he passed before us.

Yet by our passing, and by our understanding of what we are passing, we can often contribute something to the sum of man's pleasure. Either by example or by word of mouth, or by recorded observation. And this perhaps is where the value of a club journal comes in. If you have some kind of a journal that records the experiences and pleasures of your members, no matter how simple it may be in form, it will become something of value to be shared by others as well as among yourselves.

Loving the sport of fishing, and especially fly fishing, and the creatures surrounding it, the environment about it, the fly fisherman inevitably must seek the preservation and perpetuation of these values for future generations. You might resist the word, "conservation" — heaven knows I do — I'm tired of it. But just the same it's there and we can't get away from it. This is our pleasure today. It must be other people's pleasure tomorrow. It must be there for them to enjoy. I think this is of importance. I don't think we should reject it or ignore it.

The basic thing, of course, is the magic wand that we take out with us, and its tapered line, and the light leader, and the feathers which ornament the hook. It's enough to concentrate on these. If we do, and if we are true fly fishermen, the rest comes naturally because it is in the line of our tradition, and, just as naturally, it is bound to project ahead into the future of our tradition.

Acknowledgements

"The Man Behind the Rod" was originally a CBC broadcast, and has not appeared in print until now; "Fascinating Challenge"; "They Pass in the Night"; and "Watch the Creek Mouths" were previously unpublished.

With the exception of the above, the pieces in this book originally appeared in the following publications:

The American Sportsman
"Along the Steelhead Rivers"; "In Search of Trout"; "Salmon of the Vatnsdalsa"; "The Splendour of the Run."

British Columbia: a Centennial Anthology
"Diplomat's Fish"

The Creel
"A Talk to Oregon Fly Fishermen"

Field and Stream
"Big Floaters for Western Trout"; "Chilean Trout Fishing" (published originally as "How Good is Chilean Trout Fishing?").

The Fly Fisher
"Steelhead Angling Comes of Age"

The Fly Fisherman
"On the Trout Water"

Life
"Articles of Faith for Good Anglers"

Mayfair
"Ever Fish for Sebagos?"

Outdoor Life
"First Among Equals"

Roundtable
"The Evolution of a Steelhead Fly"

Sports Illustrated
"A Westerner Looks at the Beaverkill"; "Grandmother, What Sharp Teeth You Have!"; "The Quinault River."

Trout
"Freedom of Choice"; "Outdoor Ethics" (originally a speech to the B.C. Wildlife Federation, and first published in *B.C. Digest*. The present version is from *Trout*).